A PATRIOT'S PROMISE

A PATRIOT'S PROMISE

PROTECTING MY BROTHERS, FIGHTING FOR MY LIFE, AND KEEPING MY WORD

SENIOR MASTER SERGEANT

ISRAEL "DT" DEL TORO, JR. (RET.)

WITH **T. L. HEYER**

ST. MARTIN'S PRESS
NEW YORK

First published in the United States by St. Martin's Press,
an imprint of St. Martin's Publishing Group

www.stmartins.com

Designed by Omar Chapa

All insert photos are from the author's personal collection.

Library of Congress Cataloging-in-Publication Data

Names: Del Toro, Israel, Jr., 1978– author.
Title: A patriot's promise: protecting my brothers, fighting for my life, and keeping my word / Senior Master Sergeant Israel "DT" Del Toro, Jr. (Ret.); with T. L. Heyer.
Description: First edition. | New York: St. Martin's Press, 2023. | Includes index.
Identifiers: LCCN 2023009396 | ISBN 9781250283740 (hardcover) | ISBN 9781250283757 (ebook)
Subjects: LCSH: Del Toro, Israel, Jr., 1978– | Del Toro, Israel, Jr., 1978–Family. | United States. Air Force Special Operations Command—Biography. | Disabled veterans—United States—Biography. | United States. Air Force—Military life. | Families of military personnel—United States. | Athletes with disabilities—United States—Biography. | Afghan War, 2001–2021—Biography.
Classification: LCC UG626.2.D45 A3 2023 | DDC 358.40092—dc23/eng/20230228
LC record available at https://lccn.loc.gov/2023009396

Our books may be purchased in bulk for promotional, educational, or business use. Please contact your local bookseller or the Macmillan Corporate and Premium Sales Department at 1-800-221-7945, extension 5442, or by email at MacmillanSpecialMarkets@macmillan.com.

First Edition: 2023

10 9 8 7 6 5 4 3 2 1

Dedicated to my dad, Israel Del Toro—
I will continue to keep my promise.

A PATRIOT'S PROMISE

INTRODUCTION

MEET THIS HERO

Very few people leave me speechless, but when I met Israel Del Toro—"DT," I found myself completely without words. He is the most unforgettable person I have ever met, a hero among heroes. One thing that makes him so special is the fact that he doesn't know that he is. DT didn't even want us to use the word "hero" in the title of this book.

He'll tell you, "I'm just a guy who had a bad day at work."

In this memoir, you'll read about that day, when Del Toro was a special ops paratrooper on a tour in Afghanistan and his Humvee rolled over a buried roadside land mine and exploded. By the time he gave his last command, he could barely breathe. More than 80 percent of his body was severely burned, and he lost all his fingers on his left hand. Doctors gave him a grim prognosis: they said he would never walk or breathe properly on his own. He faced only a 15 percent chance of survival and the likelihood that he would spend the rest of his life on a respirator. When he woke up from a coma that lasted four months, DT wanted to die when he first saw his own face.

He almost died three times. He had no idea President Bush had come to see him and spent twenty minutes at his side. He had no idea that his wife, Carmen, had spent eight hours or more every single day at his bedside in the hospital. He worked hard, fought through the pain, and got better every day.

This man proved everyone wrong with a miraculous recovery.

Not only did he leave the hospital in record time, not only did he learn to walk and breathe on his own again, but Senior Master Sergeant Israel Del Toro, Jr., became the first airman ever to *reenlist* in the military after being deemed 100 percent disabled. He is a recipient of the Purple Heart, the Bronze Star, and the

Pat Tillman Award for Service. He has broken world records in javelin, shot put, and discus, and he won both gold and silver medals when he represented Team USA in the Invictus Games, a competition created by Prince Harry for wounded servicemen.

This man encompasses the truest meaning of some of the most powerful words: Heart. Courage. Love. Honor.

Humans are capable of heroic acts they cannot imagine, but not everyone is willing to be that hero. Everyone can turn their lives around, but not everyone is willing to be so disciplined to keep after it day after day, month after month, year after year—and then to become an advocate for others beginning the journey. Maybe everyone can do what DT has done, but not everyone would have. DT is one of the most resilient among us, and the most resilient can indeed change the world.

He went into the military because he wanted to be able to someday say to his grandchildren, "Let me tell you my story." Oh, I guarantee his great-grandchildren will hear this story. Two hundred years from now, people will be telling the story of this man. There is a pride in his family—and in our nation—for generations to come. I believe this man is on the precipice of becoming one of the greatest forces for good and change on the planet.

DT's story gives us this perspective: our dreams—and our tragedies—show up in ways we never pictured them, and the ripple effects of our lives have an impact in ways we cannot imagine. Who you're becoming and what you've done will impact far more than you know. You can change your family tree in your quiet way, and the difficulty you're going through is the very thing that will build your legend and your legacy. If you've ever wondered if you have what it takes, read DT's story for motivation, inspiration, and a dose of perspective.

Millions of people will read this book, hear this story, and nod their heads in agreement about this man, Israel Del Toro, Jr., senior master sergeant of the United States Air Force, the man who changed the world. In this book, he will finally tell his whole story, but I wanted to brag about him for just a moment before you dive in.

DT, we are proud of you and thankful for you. You did what your dad told you to do, brother. You've taken care of your family. Your impact reaches further than you know. Give yourself some daggum credit.

Ed Mylett,

global speaker, world-class entrepreneur, top-rated podcaster, bestselling author, life and business strategist, influencer on YouTube (594k) and Instagram (2.3 million)

CHAPTER 1

2005

On Thanksgiving in 2005, the White Sox had just swept the Astros in the World Series, and I had received my orders to capture or kill a high-value target in the southeast quarter of Afghanistan. Everything was coming together just as I had been hoping for a long time.

As staff sergeant in the Air Force, I was a trained observer and expert marksman on the ground, coordinating ground troops and calling air strikes in the most dangerous situations. Our mission was an air assault operation into rugged terrain behind enemy lines, a bowl-like valley tucked into the rough mountains of Zabul. We were to locate our target, capture or kill the enemy forces, and destroy the supply route the Taliban was using. Wherever the intel was—which usually meant the most danger—that is where the Air Force sent me. The hunt was on.

The first week of December, we arrived in the valley, right next to a small town with only one road. All the town's traffic came through this one road, and we traveled along it to bring in our dirt bikes and Humvees, one regular and a gun truck. We set up our base of operations in a villa in the valley, and then, with the company of the Afghan National Army, we went into the town to talk with the elders, asking if anyone had seen anything we should know about.

I was confident we would find our guy, but a couple of days had passed and we had not found our target yet. We knew we were close. Unnervingly close.

Do you know that eerie, hair-raising feeling you get when you just know someone is watching you? That is how we felt. They could see us, even though we could not see them. Our Afghan interpreter heard chatter through radios: "We see the Americans." The Taliban were watching us from the mountaintops.

We needed to get out of the bowl of the valley and seize the advantage from the mountain. We planned our strategy, dividing our scout team into two groups. One group would leave in the late afternoon, travel through the night, and explore the mountains on foot. They would stay overnight and capture the enemy as they came up the mountain in the morning. The other half of our scout team—the team I was on—would be on overwatch to observe the terrain and support the guys on foot. If anything went down, we would see it first.

I was the only noncommissioned officer; the rest were privates. My teammates and I were on a rigid schedule of two hours of sleep, one hour of watch, taking turns through the night with heat-vision glasses, watching for anything at all. It is a curious blend of tense and peaceful out there, waiting in siege under the most tranquil sky. In that total stillness, I could see every star. Just like Forrest Gump tells Jenny about the sky in Vietnam: "I couldn't tell where heaven stopped and the earth began."

Another night passed, and we did not have our guy. We had not accomplished our mission.

In the morning, our guys on overwatch went back down to the trucks, and we drove into town to meet with the Afghan National Army. I remember eating breakfast with the elders of the town. They served us goat meat and what we called "sand bread," because they knead it in a sand bowl. When you eat sand bread, you can feel grains of sand in your teeth for days.

Suddenly, our lieutenant saw movement on a small hill nearby. He spotted these guys watching us from above. This was our chance. I threw on my lucky hat, and we hustled into the trucks. Lieutenant took the first truck, and my guys and I took the second, racing after the lieutenant even before we knew what he had seen.

Our lieutenant was green with inexperience. He was a fresh graduate on his first deployment, and he had not seen much yet. This guy was so naive that when we drove past an IED on the side of the road, he took a selfie with it. I mean, come on. I was forever respectfully guiding him back on course, but I often found myself thinking, Come on, man. Don't be an idiot. Get your shit together.

I radioed the lieutenant as I was trailing him. "Sir, what are you doing?"

"DT," he said, "I saw these guys. We're behind them. I think we can capture them."

"All right, sir. What do you want to do when we get up there? Who do you want to take with you?"

"I want to take you and the interpreter."

I held the radio in my hand, and I thought, Let me get this straight.

Out of all of us, he wants to take our interpreter

and me. Our interpreter doesn't carry a weapon, and he's the only one who can understand the Taliban communication. As for me, yes, I'm an expert marksman, so I am ready to go at all times, but I am also the only fire support element who can directly communicate the positioning of weapons and fire onto our target. To take me and the interpreter is to put all the heavy hitters in one spot without much support.

I thought, I swear to God, sir. If I get shot, I'm going to shoot you right in the ass.

The mountains of Afghanistan are jagged boulders. The terrain is steep, and there are rocks everywhere. This rugged landscape is enemy territory, and the guys who live there know the terrain far better than we ever could. To find them would be a once-in-a-thousand opportunity. If we had our chance to shoot, we should take it.

I scanned the area with my rifle, and I spotted the enemy.

With my sights on him, I put my finger on the trigger, ready to take him down.

I called to Lieutenant over the radio. "LT. I have sights on the target. I am ready to shoot."

"Don't shoot," he said. "We can capture them."

I watched the enemy through the scope of my gun.

These guys might look vulnerable in their bare feet and flip-flops, but they were like gods on the mountain. They jumped all over those rocks with the nimble agility of mountain goats. They'd be gone in an instant. This might be our only chance.

I repeated, "Sir. I have them in view. I have the shot."

He repeated, "Don't shoot, DT. We will capture them."

I followed the orders. I moved my finger off the trigger.

But just as I suspected, by the time we got close enough to capture them, they were gone. We had missed our chance.

We drove back into the valley, our target still at large.

The day before, after we split our team in half, our teammates who were scouting on top of the mountain had only taken enough supplies for one night. They thought it would be enough. They radioed down to us, asking for more supplies, so we piled into our two trucks to travel again on that one road out of the village.

Normally, I would have ridden in the second truck, just like when I followed the lieutenant into the hills.

My role was always to protect the truck ahead of me, with my hand on my weapon to defend them from anything they couldn't see. Since this was a short trip, just two or three blocks to meet up with our guys, I jumped into the first vehicle with the lieutenant. I rode in the front passenger's seat, watching the road ahead.

Our Humvee drove through a shallow creek, slowing down in the middle as the water splashed up around the vehicle. The lieutenant revved the engine to pull us up on the shore, and seconds later, just a few meters past the creek, I heard a loud boom. The Humvee jolted and the ground shook beneath me. I saw a flash of white, and I felt an intense blast of heat on my left side. There was a shattering noise of broken glass, and there was flying debris all around. The air tasted like gunpowder, the smell thick and bitter.

Holy crap, I thought. We just got hit.

Let me tell you, I had heard stories of this happening. I had listened to people talk about explosions in battle, about how time slowed to a crawl, and how their lives flashed in front of them during their moments of greatest danger, but I never believed a word of it.

Until it happened to me.

When I got hit, images flashed through my mind. Picture after picture of events of my life appeared with crystal clarity. But for me, they weren't memories recalled—they were moments yet to come. I saw events I wanted to experience. I saw reasons for me to fight for my life.

I saw a vision of my wife dressed in white, of the two of us finally getting married in the church. We had been married by a justice of the peace, and we had planned three times to have our wedding in the church. But every time, I was deployed. I wanted to give my bride the wedding she longed for.

I saw a vision of the two of us honeymooning on the islands of Greece. I saw the turquoise water, the white sands, and the blue-domed buildings on the coastline. I wanted to take my wife on the honeymoon we had imagined.

And I saw a vision of my three-year-old son, of me teaching him to play baseball. I saw him holding his leather glove in the air, the ball landing in the soft leather of his palm. I wanted to have a catch with my boy. I needed to get home to him, to teach my son to play.

In that instant, something inside me shouted, Get out of this truck. *Now.*

I popped the door open and threw myself out of the vehicle, but I was already on fire from head to toe.

I recalled that creek we had driven through seconds ago. I turned to run toward the water—but the flames overtook me. I collapsed to the ground.

I lay there, burning alive.

I thought, *I broke my promise.*

I had broken the promise I made to my son, that I would never let him grow up without a dad. I had broken my promise to my wife, that I would come home to her. And I had broken my promise to my dad, that I would take care of my brothers and sisters.

I'm going to die, I thought. *I broke my promise.*

CHAPTER 2

1990

"Junior! Come inside!" my mom called to me. "Your dad is on the phone!"

I was twelve years old, dribbling a soccer ball in the street between my house and my cousins' place a few houses down in our neighborhood on the east side of Joliet. I had dreams of becoming the next Mexican soccer hero, and nothing was holding me back from playing in the FIFA World Cup. I was kicking a soccer ball every waking second of the day.

My dad and I played soccer together, it was our game. I was practicing my new moves to show off when he came back home from Mexico. He had gone to stay with his sister down in Guadalajara to get medical treatment for his lungs. He had been struggling for breath for the last couple of years, living on a respirator. We hoped the warmth of Mexico, the care of his sisters, and the experimental medicine might be the cure he needed.

I kicked the ball up the driveway and onto the porch while my mom waited with the phone against her shoulder, its long, coiled cord stretching back to the kitchen. I kicked the ball right up to the door, reaching for the phone.

My mother pulled the metal screen door closed between us. "Junior, no kicking the ball in my house. No, no."

"Okay, I know," I said, lifting the ball under my arm, unzipping my jacket. There is a special kind of sweaty for a boy who plays outside in January in Chicago. The air can be icy and gray, and your breath will blow out in puffs, but the temperature inside our jackets is another story.

I reached for the phone, eager to hear my father's voice.

"*Hola,* Papi!" I said, careful to speak in Spanish. My mother and I spoke mostly English at home, but my father would not allow his children to speak a language their grandparents couldn't understand. When I spoke to him on the phone, he required Spanish.

"Junior! My boy! How is *mi hijo* today?" I could hear his raspy breathing through the phone.

"*Estoy bien!* I'm good!"

"Your day at school?"

I told him about my classmates, my friend Paul from my basketball team. I had finally started to enjoy school, glad to transfer to the private Catholic school my neighborhood friends attended, instead of the Catholic school of my early elementary years. Still a Catholic school, but now with my buddies.

My mother was making dinner for the other children, my two sisters and my brother. Our big family has an equal pattern with the four kids—boy, girl, boy, girl—and ours is the smallest branch of our family tree. On my dad's side alone, I have ten pairs of aunts and uncles, each with at least six kids. We do family in a big, loud way.

My dad and I talked about soccer, basketball, normal father-son chatter about the things on my twelve-year-old mind. He made sure I was behaving at school, eating my vegetables at dinner, minding my mom, and keeping my grades up. All the usuals.

"Do you want to talk to anybody else here?" I asked, offering up my brother and my sisters.

"No, Junior, I talked to them already. I want to

talk to you about one more thing tonight." I heard him draw a labored breath.

He said, "*Tu eres el hombre de la casa, Junior.* You're the man of the house when I am gone, the oldest one. I need to ask you to make me a promise, Junior. *Necesito que me hagas una promesa.*"

"Sure, Papi, anything."

He fought to take in another breath, and then he said, "*Prométeme que siempre cuidarás de tus hermanos y hermanas.* Promise me you will always take care of your brother and sisters."

"What? *Sí,* yes, of course, Papi. Why would you even ask me that?"

"Promise me, Junior. *Prométeme.*"

"*Sí,* Papi."

"*Lo dices.* Say it."

I repeated the words he needed to hear. "I promise to always take care of my brother and sisters."

My brother, Angel, popped the soccer ball out from under my arm, and I jumped with the reflex to get it back, pulling the phone cord across the kitchen, breaking my mom's rules and her concentration while she cooked dinner. "Got to go, Papi. I'll talk to you tomorrow—*te hablaré mañana*—"

I tossed out my distracted goodbye as I chased my

brother and my ball. My mom scooped up the phone to finish the call with my dad.

That was the last time I heard my father's voice.

The next day, just like any normal day, my mom picked all of us up from school—my brother, my sisters, and my little cousin. As we drove into our neighborhood, I spotted two more of my cousins waiting at the driveway, the two oldest in the family, both named Luis.

"*Mis primos!* My cousins! *Luis Chico y Luis Grande!*" I called to them, so excited. What a treat that they would be there, and even my mother was surprised. My brother and sisters tumbled out of the car, all of us in a rush to be the first to get to them.

That's when I noticed the serious looks on their faces.

"*Cuál es el problema? Qué está pasando?*" I asked immediately. "What is wrong? What's the problem?"

Their greetings were reserved and quiet as they walked us inside. My cousins said, "Hey, let us talk to your mom alone for a few minutes, okay? Go on into your bedrooms."

I obeyed and went to my room, but I would not be left out of whatever would happen next. I peeked through the gap between the hinges of my open bedroom door, spying into the dining room where they spoke softly to my mom.

The whispers were suddenly broken with my mother's wail, a long, loud, high-pitched cry. I watched from my bedroom as my mom dropped to her knees, her face to the floor.

Even before I learned what had happened, I knew everything had changed.

I rushed into the dining room, asking, "What's going on? *Qué pasó?* What's happened?"

One of my cousins knelt to the floor with my mother, and the other came to me. He said, "I'm so sorry, Junior. *Tu padre ha muerto.* Your father died today."

I looked back and forth between all of them, my two cousins bearing the news, and my mom crumpled on the floor. I knelt beside her, listening to her cry. I'd never heard sounds like that come out of my mother.

In between her broken breaths, she put her hand on my arm. "Junior," she said, "take your little cousin home."

My mom's sister lived on our street, just a few houses down, and my parents were godparents to my younger cousin. To take him home meant that I would be the messenger delivering this devastating news to the rest of our family, just like my cousins had delivered the news to us.

I went into the bedroom to get my cousin, and I walked him to his home. I remember there was snow on the ground. The short walk to their house seemed like the longest mile.

We stood on the front porch and knocked. My aunt opened the door, calling to my uncle when she saw me. As soon as I said the words, "My father is gone," my aunt and uncle leapt off the porch, bolting past me to get to my house, to get to my mother.

I walked home on my own, my jacket hanging open and my snow boots dragging in the gray slush-snow. Our front door stood open, and I could see my family inside. I could hear them, the shouting frenzy of so many questions and so few answers, of so much sadness.

I walked down the alley between the houses, straight into my backyard. My father's voice was still vivid from the night before: *You're the man of the house when I'm gone.* I wondered what a man does in a moment like this. I dropped to my knees, the snow soaking into my jeans. I wrapped my arms around myself, and I wept. A child with nobody to hold him.

I learned later that my dad was at his sister's house when he started struggling to breathe. He had been to the doctor the day before—on the day I spoke to him. He was on a strict diet to strengthen his heart and his lungs, but that afternoon after the visit to the doctor, he splurged and treated himself to something he shouldn't have eaten.

I wonder how much he knew when he spoke to me that night before he died. I wonder if he knew he had risked a gamble his heart couldn't afford.

My *tía* said he called out to her from the breakfast table. When she rushed into the kitchen, he was holding his chest. *Tía* helped him to the car, and she drove my dad to the hospital, among the honking taxis and manic traffic in the streets of Mexico. She talked to him as she drove, desperately trying to keep him awake with the sound of her voice. "*Háblame, Israel.* Keep talking to me, Israel. Keep talking to me."

Suddenly, he became very still, very quiet, and his body slumped over onto her as she drove. My father died in her car, on the way to the hospital.

Our family traveled to Mexico to bury my father in the family cemetery. I had seen cemeteries in the United States, with the rows of gravestones and impeccable landscaping, but in the dry climate of Mexico, it was a waste of water to try to grow grass in that space. The cemeteries are mostly shades of gray and white, so the Mexican marigolds decorate the graveyard with splashes of orange. We buried my father in a plot of land surrounded by flowers and candles. I wore a stiff suit and shoes that hurt my feet.

When we came back home to Chicago, I hung up my cleats and put the soccer ball high on a shelf in the closet. If my dad wasn't here, I didn't want to play.

My dad was everything to me, my hero. I followed him everywhere. I did everything he did, followed his every instruction. So, I already knew I planned to keep my promise, no matter what it required of me, for the rest of my life. My dad had taught me to be a man of my word. His words rang in my mind like a resounding gong.

Promise me, Junior. Promise me you'll take care of your brother and sisters.

CHAPTER 3

1990

My dad was the compass of my mother's life. When he died so suddenly, my mother lost her way.

I didn't understand any of this when I was a child, but I can see now how her sadness affected her, sending her searching for anything that would numb the pain of her broken heart. Almost immediately, she turned to other men. My father was gone, and my mother started slipping away.

At first, she dated my father's friends, men with familiar faces and voices that we knew. Then she started dating people we didn't know at all. She brought home men of all ages, each one younger and younger, even a kid as young as eighteen years old. My father hadn't been gone for a year, and now my mom's boyfriend was only five years older than me—the same age difference between me and my littlest sister.

I didn't know how to keep my word to my dad, and I really didn't know how to bring my mother's attention back to our home. I wanted her to focus on her children, not other men.

I decided to course correct the only way I knew how: I'd tattle.

Each night, when my mom and her boyfriend slipped into her bedroom by themselves, I snuck into the kitchen. Silent as a mouse, I lifted the phone off its cradle on the wall, turned the rotary dial as quietly as I could, and called for help. I'd call my grandparents, my aunts, or my uncles—anyone in my family who would come over and interrupt those bedroom sessions. With my aunts and uncles just a few houses away, I knew they'd come over and break things up right away. My mom and her boyfriend thought they could keep their secrets, but I was determined to break their pattern. I'd call for help every single night if I had to.

One night, everyone was asleep but me when I heard the soft click of her bedroom door close. I snuck into the kitchen, took the phone off the hook, and called my grandparents and my aunt to come over and break things up once again. This time, my grandparents had had enough. They came to the door, barged straight into my mom's bedroom, and kicked this guy out of the house once more.

But this time, my mom beat the crap out of me. She chased me into my bedroom, beating me with *el cinto,* a leather belt. I cowered in my bed, raising my arms to protect my face against the blows. She whipped me again and again, shouting, "Junior, so help me, God, if you do this again, I will send you to boarding school. You'll never see your brother or sisters again. Is that what you want, Junior? I swear to God, I'll do it. You will never. See. Them. Again." She whipped me with every word.

After she was done, she left and slammed my bedroom door closed. The air buzzed with the silence after her shouting.

Shit, I thought, fighting the sting of tears, refusing to let myself cry even as my body throbbed. Now what do I do? How can I take care of my brother and sisters if I'm not even here?

I balled my fists and steeled my heart, and I made a decision that night. Maybe I could not correct the course for my mom, but I would never let anyone take me away from my brother and sisters. I would care for them any way I knew how, even if that meant letting my mother lose her way.

I hated this young punk who was hanging out in my house with my mom. I didn't want anything to do with him, and I didn't want him around any of us. I remember the day I took him on, and he hit me. His punch threw me back to the ground. I looked up at my mother to see what she would do to this dude who just hit her son, this teenage boyfriend who was beating her child. I thought for sure this would be the event to wake her up to what was going on.

But she did nothing. Her eyes were empty. There was nothing there.

She had nothing left. She just let him hit me.

That kid walked away like he had taught me a lesson, like he had won the fight.

Tell me, what's a thirteen-year-old kid supposed to do then? How's a child supposed to keep his family safe?

I focused all my energy on my sisters and my brother. I watched over them when we were together, and they were in the center of my every thought when we were apart. I stayed up late at night, making sure the house was locked up tight before I was the last one to sleep. I checked their backpacks in the morning and their homework at night. I knew where they were at any moment of the day. Someone was going to parent them, and that monumental task had fallen to me.

My mom and I could keep the peace with each other, as long as I stayed out of her relationships. Things seemed to settle down a bit, and it seemed for a while like everything might be okay. It even seemed like we were on the same team for a little while. I was finishing up eighth grade, looking forward to high school, planning the next steps of our lives as best as I could. She was taking care of herself, and I was taking care of everything else.

But I never let down my guard. A child who has lost one parent is always keenly aware of the risk of losing the other. With the death of my dad, I was forever aware of the possibility of losing my mom, too. I had a recurring dream that she died, and I couldn't get it out of my head, day or night.

One day, my mom and I were alone in the car together. With her eyes on the road and not on me, this seemed like a safe time to tell her about my dream.

"Mami," I said, "I keep thinking you're going to die."

"What?" she asked, glancing to me and then back to the road. She reached a hand to me, ruffling my hair. She said, "No, nothing like that will happen, Junior."

"But what if it does?" I asked.

"Junior, it won't," she said. Her right hand was warm on the back of my neck, her left hand resting on the steering wheel. But I needed a plan, not an empty promise.

I said, "Mami, just tell me. If you die, where do you want to be buried?"

She started to laugh, like my question was ridiculous and funny. But then she caught my eye. Her face changed when she saw that I wasn't laughing.

"Okay, Junior," she said. "I want to be buried at that cemetery with the fountain."

That's the only plan she gave me. The cemetery that had a fountain. That information would have to be enough. I didn't want her to die, but if she did, I had a plan.

Memorial Day weekend came at the end of the school year, and we were gearing up for my eighth-grade graduation weekend. At our school, the eighth-graders took a class trip to Six Flags Great America, a tradition I had looked forward to for a long time. I couldn't often set down the weight of carrying the burden of our entire family, so this day would be a great relief. A rare chance for me to be carefree, to act my age, to be a teenager.

The night before Six Flags, I woke to the sound of someone pounding on the door. I looked at my clock to see it was 2 A.M. I came out of my bedroom and peered through the peephole in the front door. I saw a badge and a uniform. It was a cop.

I opened the door a few inches to look out at him.

"Is this the home of Maria Del Toro?"

"Yes, sir."

"Are you here alone?"

"No, sir. I'm not alone." My sisters and my brother were all asleep in their beds.

"Any adults at home?"

"No, sir."

"Can you contact someone?" he asked.

"Why?"

"Son, I need you to call someone, please. Any adult you know."

He stepped inside the door, and I went to the kitchen telephone to call my aunt, waking her from a dead sleep.

"*Tía?* This man needs to talk to you," I said, handing the phone to the police officer.

He said, "Ma'am, you are the sister of Maria Del Toro? I'm sorry to tell you, ma'am, but there has been an accident."

My stomach dropped like a stone. My arms fell heavy to my side. I listened as the officer dropped the anvil on my life.

My mother had been on a motorcycle with that eighteen-year-old boyfriend who beat me. The two of them were on a powerful little bike called a crotch rocket. They had one helmet between the two of them, and my mom's boyfriend wore the helmet that night. A drunk driver cut in front of them, my mother was thrown from the bike, and she landed on her head. She had been airlifted to Loyola Medical Center. I learned all this information as the police officer told my aunt.

The following days are a blur in my mind. I was the only one of us kids old enough to enter the ICU, so I was the only one who got to see her. My mother looked so small as she lay in that big bed, her head wrapped in bandages, her face bruised and swollen. She was in a medically induced coma. Every day for several weeks, I visited her hospital room. I sat in the plastic chair next to her bed, staring at her sleeping body, listening to the beeping machines keeping her alive.

In early July, after weeks of sleep, she woke from her coma. Her eyes fluttered open. I leaned close to her, holding her hand. "Mami, it's me. It's Junior."

She had a tracheostomy tube to help her breathe, so communication was difficult. She could only whisper, "Junior, I need."

"What do you need, Mami?" What did she want? What could I give her that might help anything at all? How could I fix this?

I leaned in closer to hear her raspy whisper, ". . . M&M's."

She wanted candy? Of all the things I could give her, she wanted chocolate-covered candies? She hadn't eaten in weeks, and though I knew very little about comas and hospitals and medical protocols, I knew I couldn't give her milk chocolate. I wanted to give her anything she wanted, but she asked for one thing I couldn't give her.

"Mami, I can't," I said.

"Please, Junior," she whispered, "I'll just suck on them."

That was all she said. Seconds later, she slipped back into a coma. And I never heard her voice again.

In the following days, her entire body swelled up with fluid. They tried to save her, draining the fluid to reduce the swelling, but her organs began to shut down. Her body couldn't maintain the fight for her life. On July 15, she passed.

I was merely fourteen years old, and now I was the leader of us all. Just a year and a half after our dad died, I gathered my little brother and sisters around me, and I broke the news. "Guys, I'm sorry. We lost our mom."

We buried my mother in the cemetery with the fountain, just as she asked. Each person in my family stood by her gravestone after her funeral, but I kept my distance. When everyone had gone back to their cars, I went to her grave by myself.

It felt similar, to do all of this again so soon, in the stiff suit and the tight shoes. Yet somehow, the loss felt entirely different. It struck a different part of my core.

I stared at her name on the gravestone, quiet for a few long moments.

And then, to this day I hate to say this, but I cursed her name. I looked at her gravestone as if I were looking in her eyes, and I cursed her name.

Look what you've done to me—to us. Why?

Why were you with an eighteen-year-old kid?

Why did you get on his bike?

Why didn't you listen to me, Mami?

Why did you let this happen?

I cursed the world I was born into, and I cursed my mother's name.

I had tried to get my mom on course, but now she was gone.

I had tried to protect my brother and sisters, but now they were orphans. So was I.

I was trying to keep my promise to my dad, and it was becoming so much to carry.

To this day, I've never been back to her grave. I could never go back to the place that would remind me of all that had been broken—my family and my promise.

I cursed my mother's name. But in my soul, I felt like I was the one who was cursed.

CHAPTER 4

With both of our parents gone, Grandpa and Grandma became our guardians. This was a culture shock all its own. The generational gaps brought a difference in every opinion: beliefs, politics, values, and—most painfully—gender roles. Grandpa believed men don't cry, but women do. Men wear pants, and women wear skirts. Boys play outside, and girls play with dolls. He believed a young woman's place is in the kitchen, learning to cook for her future family. He felt that a young woman—specifically, my littlest sister—had no place on a ball field. Zero sports, ever.

"*No. Absolutamente no,*" my grandpa said, pounding his fist on the table. He was very old-fashioned, and he was in charge of us now. "No girl *en mi familia* will ever play sports. *No fútbol, no béisbol, no deportes.*"

All four of us kids had been athletes, and my father had played ball with each of us before he died. But Grandpa was from the other side of our family tree, my mom's father. He raised girls differently. After all that we had lost, now my sister wasn't allowed to play sports anymore. The tension in our home was crazy.

I had to be a go-between for my little sister with my grandpa, to somehow help them understand each other. It was a great gap to stand in, and sometimes I felt like the two of them could pull my arms right off as they pulled in their opposite directions.

School was another kind of tension all its own. I had always wanted to go to the University of Illinois, and my mom and dad had encouraged me to chase after that dream. Now, with my parents gone, the school counselor discouraged me from that goal, suggesting instead that I go to a community college. He said something about it being more affordable, close to home, and in a supportive environment. He tried to make it sound like it was in my best interests, but I knew what he wasn't saying. He did not think I could handle the pressures of a four-year university.

If you ever want to motivate me to do something, tell me you think I can't.

I applied to U of I, and I got a full academic scholarship. The morning after I got that piece of mail, I marched into school and slammed the acceptance letter on my counselor's desk. I said, "You can eff off, man. I'm going to U of I."

That moment of victory cost me a month of detention. Completely worth it.

I was finally heading off to school to build a life of my own. That first semester of college was just what I needed—a rush of new people, places, faces, events, experiences, and independence. The University of Illinois was only two hours away, but it was far enough. I spent weeks away at a time, finally free from responsibilities far beyond me.

But I was still the peacemaker in our family, and I needed to check the temperatures in the relationships. After many weeks away, I came back home to visit. I remember our family was sitting at the dinner table when we heard a loud crash from the bathroom. I raced down the hall to find that my grandfather had fallen. He was a big man, and the task of lifting him required all the strength I had. I don't even know how any of my family could have helped him if I hadn't been there.

We got him to the hospital, and we learned the cause of the fall: he had had a stroke. From that day forward, he was paralyzed on the left side of his body.

I already wondered how my family navigated life together while I was away at school, but now that my grandpa was paralyzed, what would they do without me? My grandma would be the only adult in our home, continuing to take care of my teenage brother and sisters, and now my grandfather, too. How could I leave them now? That question dove to a deeper level when I thought about going back to school.

I decided that I would drop out. That was all there was to it.

I told my grandmother my plan to stay home, but she wouldn't allow it. When my family feels strongly about something, we hear about it. She knew the University of Illinois was my dream. As adamant as my grandfather had been about my sister not playing sports, Grandma was just as adamant that I would not quit school. "Go, Junior," she said. "You must go."

I went back my sophomore year with great hopes to stay, only to receive a call a few weeks later with more bad news: my grandma was diagnosed with pancreatic cancer. Even though she wanted me to stay in school, neither of us could ignore the pull for me to come home.

Always and forever, my promise to my dad echoed in my soul. So I gave up my full-ride scholarship to take care of my family. I didn't even finish the end of my sophomore year.

That's the first time that I recall noticing the pattern of the curse. It's the same feeling I felt when my mother died, and it became a pattern in my life: anytime anything good happened for me, something terrible was right behind it.

When I moved home to take care of my family, I was shocked by the tight finances. Where had the money from our parents' estate gone? What happened to the settlement from suing the drunk driver who killed my mother? I did not expect that everything would be taken care of for the rest of our lives, but I knew there had been money to sustain my family for a while. Longer than this. Where had everything gone?

I began looking into our family finances, and I discovered the worst news yet. The money had evaporated into a void of debt created by my grandparents, aunts, and uncles. They were assigned to our care, and they mismanaged the money allocated to us four children.

I hated to do it, but in the name of integrity, I had to find out where all the money had gone. I hired an attorney, and I sued my grandparents, my aunt, and my uncle. After a lawsuit against our extended family, my siblings and I finally received what little was left. We each had ten thousand dollars, and we had each other. That's all.

Consider the annual expenses of a family of four, calculate how far this money might take us, and you'll know what I did next.

I found a job working at a chemical plant, and I stepped in as the guardian, support, and leader for our family. I was twenty years old, and I was responsible for every decision—maintaining our home, making doctors' appointments, transporting my brother and sisters anywhere they needed to be, and signing permission slips and checks for everything they needed. Their safety, well-being, and financial stability all rested on me.

We kept that system in place for a couple of years. My paycheck brought in good money to support my brother and my sisters. I was keeping my promises and meeting the expectations.

I felt bored and a little suffocated by this life I'd been given, but that seemed beside the point, really. Sometimes you must do what needs to be done, whether it challenges you or not. Sometimes, it's what a man must do for his family.

On a rare day off, I was chilling on the couch and watching TV when I saw a commercial for the Air Force. I'd probably seen it before, this collection of scenes of airmen in action, their jets taking off into the sky. The commercial showed all these people working together, calling air strikes and jumping out of planes, and they all looked like Rambo. Their tagline read, "Aim high," and then they listed a toll-free number to call a recruiting center right here in Illinois.

My first thought was, Nobody in my family has ever been in the military.

My second thought was, Why hasn't anybody in my family ever been in the military?

And then I started doing the math. I was twenty-one. My sister was nineteen and engaged to be married. My brother was seventeen, and my youngest sister was fifteen. They all still had some growing up to do, but everybody was old enough for me to step away. Maybe it was time. Maybe it was *my* time. Time to write my next chapter, to do something with my life. Nobody in my family had been in the military, and maybe it was time to change the script.

I drove to the recruiting center, and I went right past counters for the Army, Navy, and Marines. I had heard that the Air Force had the best benefits and the prettiest girls, so I made a beeline for the Air Force counter. I asked a few questions, answered a lot more, signed my name, and agreed to the terms of service. A few weeks later, I headed off to basic training at Lackland Air Force Base, just ten miles outside of downtown San Antonio, Texas.

And that's how you say, "Let's go."

CHAPTER 5

Some people say, "Air Force? Pfft . . . more like *Chair Force*. Basic training is a piece of cake."

I call them "people who don't know what they're talking about," because that's one of the stupidest things a person can say. A comment like that only shows their ignorance, literally that they don't know what they don't know. They are *wrong*. Dead wrong.

Air Force boot camp is the military's seven-and-a-half-week training program, and every airman has to graduate before he or she can join the Air Force. About thirty-five thousand trainees graduate each year, and the average age for a trainee is twenty.

Right away, trainees are assigned a smaller squadron called their "flight." These people become their teammates, whom they will quickly learn to protect, no matter the cost. Trainees learn from military instructors how to line up and march in formation, and the mornings begin with physical training before the sun comes up. They train doing sit-ups, push-ups, and running, all preparing for the fitness test in the seventh week.

Trainees learn the rules of the dormitory, where they have to keep their living areas and their uniforms in pristine condition. Both are inspected regularly, down to one loose thread. If a trainee doesn't meet the standard, he or she will be expected to make immediate corrections. The process may look polished and precise when you see it outside basic training, but it's not an easy learning curve for everyone. Some guys struggle to fall in line, and it can take a couple of weeks before it looks good.

Sure, there's a lot of yelling, a lot of shouting, a lot of correction when trainees get it wrong. But listen, it wasn't that long ago that I was getting hit by my mom's boyfriend or beaten with *el cinto.* Compared to that, I didn't mind the yelling at all.

During the first couple of weeks of training, there is so much to learn and so much to know. Trainees learn drills, with a lot of precision and discipline incorporated into very detailed movements in a formation.

Even mealtimes—"chow"—are packed with details. It's the responsibility of the chow runner to announce their flight's arrival to the dining facility, all with the precise words and steps and drill movements. Trainees must master every detail, even when it seems like a mundane task, because every small task is training for the greater tasks ahead.

Each trainee is assigned a rifle to care for. They learn combat arms training and maintenance, and how to handle, load, and fire a weapon in a variety of shooting positions. Trainees practice in the classroom first, and then they move out to the shooting range to practice firing weapons. They fire a total of seventy-six rounds, and they must hit the target on at least twelve of them. Trainees who score twenty-two or higher qualify as "expert," and it all begins with fundamentals. I qualified as an expert.

In CBRNE training, trainees wear gas masks as they learn how to protect themselves during a chemical, biological, radiological, nuclear, or explosive attack. They run through a simulated attack, and then conduct post-attack reconnaissance of the area, checking for damage or unexploded explosives. The gear adds about fifteen degrees to whatever the weather temperature is, so it can be raging hot.

Trainees are also trained in tactical casualty combat care, or TCCC. This teaches trainees how to perform lifesaving first aid in combat. It's information overload, every minute of the day. They learn how to apply a tourniquet and how to evacuate a wounded airman to safety. An airman can find themselves in any location around the world and must be able to survive any situation, so trainees have to learn and maintain qualifications at all times. Some of the trainees had never left home before, they had no work experience, and this was their first time for everything. But it doesn't matter what you know or don't know before you come. It matters that you learn it now, and you learn it fast. You might enter with no confidence, but you'll leave with your share.

In week six, trainees have an event known as BEAST: basic expeditionary airman skills training. It's a culmination of all the training they have received in the first five weeks. Trainees live in a simulated combat environment where they practice everything they've learned and implement it on the range. The goal of that five days comes down to one word: pain.

During BEAST week, trainees go on a simulated mission to infiltrate a village and rescue an injured airman located somewhere inside. With the sounds of gunfire and explosives, along with the presence of role players acting as opposing forces or innocent bystanders, the trainees must be effective and efficient in a stressful environment. The trainers and role players make it as real as can be, and they show you how quickly your plans can fail you. Trainees have to learn how to adapt on the fly in any scenario, when to use deadly force, and when to absolutely not use deadly force. After all, if you shoot somebody out in the desert, and they did not have opportunity, capability, and intent to kill you, then you'll be pulled off the range and put on GS-11 work, which is a white-collar position for the government. It means you'll be at a desk.

BEAST is the toughest week, and it will show you if you're ready. Not long after trainees leave basic training, they will each be given a job and a task—and be expected to perform it efficiently and effectively, with no supervision. This rite of passage brings a sense of accomplishment, and there's a good reason why friends and family come together to celebrate the airmen's graduation. Because it's a big deal.

Just seven weeks can make such a remarkable difference in a person's life. At graduation, airmen walk a different way, talk a different way, and they exude clear confidence. Airmen have learned right away what is required, what it takes, and whether they have it. It's amazing how much a person can learn—and change—in just seven weeks.

When I was in basic training, the TACP recruiter—tactical air control party, that's my control team—came to me. He looked like Rambo, so he had my attention. I grew up watching Rambo, wanting to be that guy. Long hair, bandana, muscle-head—all of it. That's who I wanted to be. He was all badged out—ranger, paratrooper, air assault, and a beret. He started telling us what we could do in this job as a TACP: jump out of planes, go to halo school, become a sniper, call in the air strikes. He said, "When you're a TACP, you are the man on the ground."

Yes, all the jobs in the military are important, but I was drawn like a magnet to the highest risk. Remember: I wanted to be Rambo.

I got selected, and I headed off to tech school. Tech school is different from basic training, because it's more of a specific learning environment with classroom studies and field training. Cadets are learning advanced training that is unique to their specialty code. I progressed through the phases, graduated, got my beret, and received my first permanent duty assignment in March 1998 at Fort Bragg. I was ready to start the next chapter in my journey as a cadet in the United States Air Force.

An airman must be prepared to survive, evade, resist, and escape any situation, so we train in some of the most remote and hostile environments on earth. I remember when we were training up in Washington State, we were learning navigation and survival in an arctic environment. This course focused on finding and killing wildlife, procuring and purifying water, constructing a thermal shelter, building a fire, and finding and eating plants to survive.

We were learning day and night navigation techniques, so we had to hike overnight through the snow. One of my teammates was from a warmer area, so he didn't know snow—let alone know how to hike in it overnight. I could tell he was getting weary, losing his spirit. I decided when we set up camp next, I would think of a way to lift his spirit again, find a way to cheer him up.

We had to catch and kill anything we wanted to eat on those survival trainings, so we learned how to set up a snare to catch wildlife. We caught a rabbit that evening, skinned him, and cooked him up for dinner. As we sat around the fire that night, I got an idea for what I could do to make my teammate laugh.

I slipped my hand inside that rabbit carcass, and I did a little puppet show. I even did a little rabbit voice and everything. It sounds ridiculous, and it was. But you do whatever you have to do to keep your teammate going.

Everybody around that fire thought I was crazy that night, but I made them laugh, and we made it through that difficult training. Maybe not everybody would put a dead rabbit's carcass on their hand and do a late-night comedy routine in the wilderness, but I would do anything to help my brothers in the field. We all made it through.

While I was in tech school, I met my best friend, Jeff Mariano. Jeff and I became roommates, and we did everything together, usually up to some kind of shenanigans in my Jeep. That year, the ladies of the R and B group TLC released the hip-hop song "No Scrubs," and everybody was singing the famous lyric, "No, I don't want no scrub, a scrub is a guy that can't get no love from me . . ." I always told him that was his song. The Jeep was mine, and that was his song. Jeff was my boy, and we were in this thing together.

By October 1999, I was on a great high and living my best life. I had completed my qualifications for all the roles I dreamt of. I was certified to call in the air strikes, and I became a fully qualified JTAC, joint terminal attack controller. I chose a path to special operations, jumping out of planes, calling air strikes, and acquiring all the skills to direct combat aircraft engaged in close air support.

When you spend your days learning survival skills, you spend your evenings blowing off steam. During our time at Fort Bragg, Jeff, our teammates, and I enjoyed many evenings at a bar called Broad Street, where Thursday nights were dollar beers and karaoke. My guys and I never took the mics at the start of the night, but fill us up with some dollar beers, and we became the princes of the place.

They called us the Backstreet Boys, all of us together, guys from all different areas. We had a pretty boy, a Hispanic guy, an Asian guy—we had everything, and we owned that place. You have to picture it, all of us singing "I Want It That Way," bringing the place to their feet with "You are my fire, The one desire . . ."

We were the best boy band in all of Fort Bragg. Not that anybody was giving out awards that year, but if they had, we would have killed it.

In every single way, I was living the dream. I had certification, qualifications, and teammates. I was a cadet in the United States Air Force, and I was prepared to defend this nation. I set a goal, and I had attained it. I was living my best life.

But in keeping with the pattern of the curse, a terrible low was to come. My grandmother died that year, losing her battle with pancreatic cancer. God, I loved that woman. Yes, we had our challenges as a family, including the disputes over custody and finances—all of that. But she was Grandma. And it broke my heart to know she was gone from this earth.

My cousin Jose died soon after, and those back-to-back losses felt like a double whammy. Jose was close to me, the one who picked me up from the airport any time I came home. The holidays were approaching, and I just couldn't bear the idea of going home. My sisters and my brother didn't need me to visit, and there seemed to be nothing to look forward to with Jose gone. I didn't come home to Chicago for a whole year. I just knew my home wouldn't feel like home.

When Thanksgiving arrived, I decided to break with tradition. Instead of going back to the big city of Chicago, I would go to my family's ranch in Mexico. I hadn't been there in a decade, not since we buried my father when I was a boy. I had become a man since then, and—it turns out—the young ladies of the town had grown up, too.

My life was about to change.

CHAPTER 6

In my family's town in Mexico, there are two kinds of guys: the real chunky guys, and skinny guys with a paunch. I called them "skinny-fat." I don't mind telling you—in that season of my life, I was in great shape. I was twenty-four years old, working out all the time, and I had officially become Rambo. When Rambo hits the street of a small Mexican town where the men are all chunky or skinny-fat, the women start to talk.

I showed up as the new guy in town. I went running on the dusty streets every morning, and the girls started watching. I saw them looking at me, whispering, giggling like they do. I mean, it's a small town. Everybody knew everybody—and their business. The girls wanted my attention, with their short skirts, their high heels, and their fake lashes, but I just jogged on by. They were all too young for me anyway.

But there was one who turned my head—probably because she didn't mean to. While the other girls were all dolled up, this one was wearing overalls that hung loose on her small frame, and her face had a natural beauty that caught my eye. She had a beauty all her own, and she had my attention. I learned her name: Carmen.

Every year, there is a festival in that town. It is the main event of the year, a carnival with colorful costumes, so much Mexican food, and a beauty pageant. The contestants compete with their costumes and talents, and the town crowns a queen and princesses. Each of the young ladies gets an escort for the pageant, and they recruit young men to usher the women in the show. They invite all the guys in town—the chubbies, the skinny-fats, and this year, a Rambo.

Tons of my dad's family live in that town, and they all call me Junior—the young version of my dad. I took my niece and nephew to the carnival the next day, so I was this young guy, walking around with two little kids, just two and three years old. We got a balloon, we had some snacks, and we were strolling the little plaza to see what we could see. Suddenly, I hear this whistling.

Let me tell you, it was not the carefree whistle of a person wandering through a market. It was the very intentional whistle of women trying to get a man's attention.

I looked around, trying to see where it was coming from, and there they were—this gaggle of women, with Carmen in the middle of them. I remember thinking, Are these women seriously whistling at me while I'm with my niece and nephew? Have they no shame?

Not that I minded, to be clear.

I made my way over to them, all cool and casual. But when I got close enough to talk to them, all those girls got really quiet. All of that whistling, and now nobody had anything to say? They just ignored me. It seemed like nobody really wanted to talk to me, they just wanted to giggle and whistle at men walking by.

Okay, girls, I thought. I can take a hint. I'm only here for a few days anyway. I don't need to spend that time kissing up to some pretty faces who all have boyfriends that will want to fight me before I leave town. No thanks, I'm out. *Adiós.*

I walked my niece and nephew to their home

around the corner, and I was content to stay home for the evening. But the girls were not content for me to stay home for the evening. They just wanted me to drop off the little children, to free me up for a good time.

Carmen came around to the house, knocking on the door. This beautiful woman had stepped away from the crowd of all those snobby girls—she wanted to have an actual conversation, and she invited me to join in the fun. I remember hearing her voice calling out my name. "Hey, Junior, you should come back out with us," she said.

I would have followed her anywhere that night.

We met on a Sunday, we talked all day on Monday and Tuesday, and then I needed to leave town to get back to work. We shared a few kisses, a lot of conversation, and a whole lot of tacos. I left soon after for a deployment in Bosnia, but I left a piece of my heart in Mexico that Thanksgiving weekend.

Carmen and I stayed in touch after I left Mexico, setting up phone calls, sending letters, and emailing almost every day. I visited her a few times, and pretty soon, I felt sure she was the one I'd like to spend the rest of my life with. I quietly spoke with her dad and her brother to ask permission to marry her, and then in November 2001, a year after I met her, I came back to town for the annual festival and pageant. There were more kisses, more conversation, and a lot more tacos—and this time, a diamond ring.

I proposed with tacos. Yes, I put a diamond ring right there in her pile of lettuce, and I got down on one knee. Don't let anyone tell you I'm not a romantic.

Since Carmen was a citizen of Mexico and I was a citizen of the United States, there was a lot of paperwork required for us to get married. Her parents had been granted lifetime visas, so we thought that might give her a foot in the door with the United States, but no. We were in for the whole, long process of passports, birth certificates, letters to the national institute of immigration saying there's no legal reason we can't be married. It was a process.

We had to have bloodwork done, and I remember that the clinic in the public hospital was a nasty situation. The medic didn't put gloves on, didn't swab the skin with alcohol, and just put the needle in and drew the blood. There was blood on the floor and the place was swarming with flies. It was not like any hospital I'd been to, and I remember thinking, Well, I know one thing is for sure. If we ever have a baby, I can't let him be born here. No, sir.

We needed to have a civil ceremony—to be married by the law—in order to start the process of getting Carmen's fiancée visa. With a ring on her finger, I promised Carmen we would have a proper wedding one day—a big ceremony in the church, with everyone there. But for now, we just needed a marriage certificate to start the paperwork. The whole thing was a long, drawn-out process. You really have to want to get married. (There was no doubt in my mind. I really wanted to marry this woman.)

We were married in February 2002 by the justice of the peace, who happened to be Carmen's aunt. Most couples have to get married at the courthouse, but Carmen's aunt really liked me, and she was willing to make an exception. As the magistrate, she could bring her gavel anywhere. We had a small ceremony, a little get-together in a garden with friends and family, drinks and appetizers, and the legal stamp of approval from her aunt. Officially, we were married.

Even with the certificate, Carmen had to go to the Mexican consulate many times to apply for a visa. It's a very extensive process for a young person, because the US government is worried the young people will come to the States and stay illegally. We had no interest in breaking the law; we wanted to do everything by the book, and that meant it took some time. Carmen was denied many times. I went back to the States for more training, and then I was deployed downrange in Iraq. Meanwhile, Carmen stayed in Mexico to jump through all the hoops required to marry an American.

Several weeks later, we got a huge surprise when we learned she was pregnant. A beautiful wife was part of my plan—yes. But a baby? Hold on.

I felt I was in the prime of my career, and I hadn't planned on us having a baby for at least another year. The timing didn't work with the schedule of my life. I can't believe this was ever true, but I just wasn't ready to be a dad back then.

I went with her to the doctor, but I was emotionally detached . . . until I saw that blinking white light on the ultrasound screen—the heartbeat of my child. The grainy images showed the newly formed brain, the curved spine, and the beating heart of our baby. Our child.

I took Carmen's hand, and I said, "I changed my mind, Carmen. I want to be a dad."

She looked at me as she lay back on the hospital bed. "I'm glad to hear that, Junior," she said, "because we're having a baby."

For years, I felt like that curse of death had followed

me all my life, like I had lost almost everyone who mattered to me. But now, quite unexpectedly, a new life had found me. The heartbeat on that screen sparked a new sense of life within me.

I was going to be a dad.

Through her pregnancy I went back to my training, knowing the baby was due in August. We wanted to know if we were having a boy or a girl, but for some reason, the baby got real modest every time the doctor tried to look around and see with another ultrasound. It was starting to look like that secret would have to wait until the baby was born.

On a Friday night at the end of June, Carmen called me from Mexico. I was hanging out with my guys, playing cards and drinking a few beers at my apartment. She said, "I have some news for you."

"Yeah? What is it?"

"Well." She paused. "Do you want to know if it's a boy or a girl?"

"Sure, of course I want to know," I said.

"We're having a girl," she said.

"Ah, okay," I answered, sipping my beer. "That's good."

"Aren't you happy?" she asked.

"Sure, yes. Of course, of course I am happy," I said.

"You don't sound happy," she said.

"Boy or girl, either is fine. I just want our baby to be healthy."

"You don't want a girl?"

"Listen, babe. I'm happy, I'm just, you know . . . hanging out with the guys. I'll be home in a couple of weeks, and we'll have our baby girl."

"Well, then," she said, "before you hang up, you might like to know one more thing . . . we are actually having a boy."

I dropped the phone and shouted, "It's a BOY!" And the place went wild.

When Carmen tells this story, she says the only sound she heard for the next two minutes was the crash of the phone hitting the floor and then a thunder of men cheering. This was before iPhones, so she couldn't see what was going on at my place—but it was mayhem. I just left my girl waiting on the line to listen to my buddies and me whoop and holler, "It's a boy! It's a boy!"

When I came back to the phone a few minutes later, I was breathless with excitement, shouting to her over the din of my guys. I said, "Carmen, it's a boy!

A baby boy! Ah, I love my baby boy! I need to go and celebrate!"

She teased me, "Sure you do. And how come you didn't act like that when it was a girl?" I could hear the smile in her voice.

Busted. A healthy baby was a wonderful thing, but she knew all along what I was hoping for—a legacy for my name, for my father's name. In my deepest heart, I wanted a boy.

There was a lot of drinking that night, and I spent a lot of money buying rounds for all my guys. It was worth every penny, every drop of celebration. My teammates and I got so drunk that night, and the Backstreet Boys made an appearance at the bar. I had never meant it more when I sang about the news of this baby boy: "Believe when I say, I want it that way."

I came home to Mexico a couple of weeks later, in time for the due date. Mexican summers are hot. Our homes are made with cinder blocks to keep everything as cool as possible but, listen, there's only so much you can do without air-conditioning. It's still *hot.* Carmen was so beautifully pregnant, and so swollen. Freaked me out a little bit, actually. I hadn't been there for the progression of that baby belly. She was, shall we say, great with child. It was game time.

My son, Israel Del Toro III, was born August 21, 2002. I will never forget his big, blue eyes—wide awake and ready for the world.

They cleaned him off and handed him to me, and together Carmen and I stared at his blue eyes and fair skin. Many of our relatives had light eyes and fair skin. In Spanish, the word is *guero.* We had thought that his nickname might be *Dorito,* since he was "a little Del Toro." But we took one look at this fair-skinned baby boy, and we knew what to call him: *Guero.*

On the day he was born, I held my son in my arms, and I whispered a promise I intended to keep: "I will always be here for you, Guero."

I was lucky to be home for a full month to help Carmen and the baby get settled, and there was more paperwork to do for a child born abroad. We had to register him with the embassy, and of course I wanted to show him off all over town. I was the proudest father since Mufasa in *The Lion King,* carrying my baby up and down the streets, holding him high for everyone to see. Within the first week of his life, everybody in the whole town knew Guero.

I covered him in a thousand kisses in that first week of his life. Each night, I rocked him to sleep and whispered my promise again, "I'll always be here for you, Guero. Always."

CHAPTER 7

As a husband and father, I continued to serve—and love—my role in the military. When I was stationed in Iraq and then Korea for a year, my missions were solo, so I couldn't take my wife and baby with me. Once Carmen got her visa, she and Guero came to the States. Carmen and Guero split their time by spending half the year in Chicago with my family and the other half in Fort Collins with her brother. She took some classes in English as a Second Language at the community college, eager to learn the language, the customs, and the life inside the United States.

I scheduled my mid-tour visits during the month of August so I would never miss Guero's birthday. That was nonnegotiable for me. I remember flying through the night to get home from Korea in time, and even after that long flight, I had to drive eleven hours to get home the night before his birthday. But I never missed my boy's birthday.

In February 2005, I got my orders to move to Italy, along with an additional piece of good news: this time, I got to take my family with me. Carmen was happy to go. The language barriers were difficult for her in the United States, but she was able to make a life for herself and Guero in Italy. She had a great circle of friends with the other military wives. They had children the same age as Guero, and it made all the difference for her to have a sense of community.

In the United States, everything was still available in Spanish. She had taken ESL classes, but she didn't have to commit to total immersion in the language. She could watch Spanish television, shop at the Mexican grocery store, and basically maintain her Spanish-speaking life without adapting to English.

In Italy, nobody spoke Spanish. Nobody. No more safety net of Spanish as an option. If she wanted to communicate, she'd have to learn English. Thankfully, Spanish and Italian are two of the Romance languages, and those similar language roots helped to bridge the learning gap from Spanish to English.

As for Guero, he was learning three languages at once—sort of a Spanglish-Italian mix of his own. Carmen and Guero watched a lot of movies together during the day, and she always turned on the Spanish subtitles so she could learn while he watched his favorite movies on repeat. His most-watched movies were *Shrek* and *The Incredibles,* and one day they watched *The Incredibles* eight times in a row. When Carmen heard a word she didn't know and the subtitles couldn't help her, she looked it up in her dictionary. This is how she began to firm up her understanding of the language.

Never underestimate the teaching powers of *Shrek* and *The Incredibles.*

The architecture of the town was very similar to her village in Mexico, and our home was about ten minutes from the base, so she could find her way around. Carmen and Guero went on walks every day—to the park, the grocery store, and even to the pastry shop. We got a dog, Dakota, a beautiful golden retriever. Carmen seemed to have finally found a home away from her home.

We had just settled our lives in Italy, and now I had to tell her the truth: I had orders to go to Afghanistan.

I can't tell you how much I really did not want to

break the news to Carmen. We had been married for only three years, and in that time, we had spent almost no time together. I had been to Iraq and then Korea, and the distance and the risks were hard on my family. Guero would get very attached when I was home, but then our bond would break during my long stretches away. We had to start over every time I came home, as my little boy even forgot what I looked like. I didn't want to leave either of them, and I didn't even want to broach the subject.

An airman learns to be resourceful with what he has, no matter where in the world he is, and our proximity to Rome seemed like something I should utilize. I took her to the Vatican to break the news. Nobody can get so mad at you when they're near the pope, right?

Yeah, that's what I thought.

Let me tell you—just know, if you ever plan a stunt like that, eventually your wife will be home alone with you. And the pope will be nowhere nearby.

When we got home from Rome, Carmen gave me an ultimatum: I could go on this tour, but then I had to get out of the military. This would be my last tour, or she would take Guero and go back to Mexico.

I said, "Carmen, you don't understand."

And she said, "No, Junior. *You* don't understand. Guero didn't recognize you when you came home from Korea, and now you are leaving again. And even though you're in the same country with us right now, you're preparing to leave, training all the time. I am raising our son alone. You said you don't want your son to grow up without you, but don't you see that it is already happening? You're alive, but you're not here. Stay with us, or we're finished."

I wanted to disagree, but I knew Carmen was right. My son would be three years old soon, and I had only been with him for a scattered total of nine months of his life. He was growing up without me.

Still, we had to complete this assignment. We decided to table that conversation until I got home from Afghanistan. No matter what happened next, I had orders to go.

All the other military wives had gone back to the States when their husbands were deployed, and Carmen didn't want to stay in Italy alone. We made plans for her to travel back to Mexico at the end of July. I would meet up with her for Guero's third birthday in August, and then I would deploy a few days later. She was so sad to go, to leave the town she loved in

Italy, the home she loved, and even our dog, Dakota. I promised her it would only be for a few months, and we would come back for it all—we would come back *to* it all. We just had to get through this assignment, and then together we would decide where to begin the next chapter of our future, with or without the military.

There was one small hiccup in that plan to get her back home: Carmen's fiancée visa had expired, and her residency card had gotten lost in all of our relocations. We had lived in four different places and multiple countries, and her residency card was nowhere to be found.

She was very concerned to fly back to the United States without a visa or residency card, but I thought it would be fine. She had a layover in Amsterdam on the way to the United States, but she didn't even need to get off the plane. I called and spoke with the consulate. They promised me there was no issue. I was sure it would be fine.

It was not fine.

When her plane landed in Amsterdam, the authorities made her and Guero get off the plane because she didn't have an American visa. She didn't have a cell phone, she didn't have any credit cards, and she didn't speak a word of Dutch. She was stranded in Amsterdam with our baby, his little backpack, and a small amount of cash. They kept her big suitcase on the plane and sent it on to America and further on to Mexico.

She spent forty hours in Amsterdam, lost and alone.

A word of advice to you: if your wife is already angry to leave her beautiful life in Italy, don't let her get stranded in another country on her way home. It's not a good situation.

She called me from a stranger's phone, and when I answered the call, she couldn't speak for crying. I told her to stay at the airport, and do not leave for any reason. She was a beautiful woman, with a baby boy, alone in a city that had a dark reputation. I couldn't get to her, and it was killing me to know how scared she was, how unsafe they were. The airport hotel was full, and she and Guero needed a place to stay for the night.

Carmen and I had to depend on the kindness of the strangers around her, and I feared that kindness was unlikely. I was afraid for her to trust anybody at all. There was one kind man on her flight, also stranded in Amsterdam, also trying to get to Guadalajara. They both spoke Spanish, and they both needed a plan until the next flight to Mexico. They traveled together to the hotel, and I told her—you get to your room, you lock the door, and you call me.

She called me from her hotel room, probably the tiniest hotel room in all of Amsterdam: a little bathroom, a tiny shower, and a twin-size bed. She locked every bolt on the door, and she stayed in for the night. She shared a bed with the baby, and she washed her underwear in the sink to wear again the next day.

(I suggested she just turn them inside out. She did not laugh at my joke. It was neither the time, nor the place.)

By the kindness of the ticket agent at the counter in Amsterdam, Carmen's travel plans were rerouted directly to Mexico City and on to Guadalajara, with no stops in the States. Still, it was an unexpected stay in a city she didn't know, and by the time Carmen and Guero came back to the airport and boarded the plane to Mexico the next day, they had not slept at all. They flew twelve hours from Amsterdam to Guadalajara. The batteries in the DVD player had run out of power, so there was no more *Shrek* or *The Incredibles* for Guero. He had consumed all the milk, all the juice, and all the snacks. He was over it all, and so was she.

They arrived in Guadalajara after midnight, and she was facing a two-hour drive to her home in Mexico, but the airport staff would not help her find her luggage—which had arrived two days earlier, without her. They said the luggage counter was closed for the night, but they would help her the next day.

I'll tell you this: my wife has a lot of patience, but she was *done.* I am married to a strong woman, and she showed her strength that day. The airport employees picked the wrong lady to cross. Do not ever underestimate the fortitude of a woman at the end of her rope, traveling through foreign countries with a hungry kid and a DVD player that didn't work.

I had to rearrange and negotiate quite a few things to make it to Mexico for Guero's third birthday. My teammates and I needed to train in the US before our deployment to Afghanistan, and I was able to finish my training early, fly to Mexico for his birthday, and then meet up with my team in Houston, where we would leave for Afghanistan. They had all my equipment, and I slipped away for the weekend with only my personal belongings and enough clothes for a couple of days. It was almost time to go downrange, but I would not miss that birthday.

I threw an outrageous birthday party for Guero. We had a huge crowd and a live band—a celebration to top all celebrations. All of our family and friends were there, and we had every reason to celebrate. Lots of food, hundreds of people, and a giant venue with a stage and lights.

But there was a heaviness in my soul when I thought about leaving in a few days. I couldn't explain it. I just didn't feel right about this deployment.

That evening, when my uncle congratulated me on throwing such a tremendous party, I told him, "I don't have a good feeling, Tío. I feel like something bad is going to happen. For the first time in my career, I'm afraid to go. I feel like something might go wrong while I'm away. Between you and me, Tío, that's why I threw this huge party for my son. Just in case, I want him to remember."

Carmen and Guero went to the airport to say goodbye to me a couple of days later, and it was the hardest goodbye of our lives.

I held Carmen close, and I whispered in her ear, "Gueris, you stay safe. Remember that I promise to come back to you. If anything bad happens, anything at all, you'll hear from me or my teammates. If I get injured, they will call you."

She looked up at me, her eyes filled with tears.

"Junior," she said, "what if it's . . . worse? What if there is more than an injury?"

"Then you will hear from my commander and a priest. If you hear from my commander and a priest, then know that I'm with my mom and dad."

She buried her face into my chest. I held her in my arms. I kissed the top of her head. "I promise to come home, Carmen."

And then I had one last goodbye to say—to my boy. Guero wrapped himself around me. He wouldn't let go. I gave him every promise in the world, but I have often wondered if somehow he knew. Things were about to change.

CHAPTER 8

Back then we didn't have the technology that we have now, so I couldn't call Carmen every day while I was away. But we did have emails—which were better than the snail mail. I emailed her as often as I could.

In the military, we are trained to beware of the possibility that any correspondence could be read at any time. The server could be hacked or the email could be intercepted by the wrong hands, so we used a code language that we had arranged with our families before we left.

Some guys will tell their wives, "I'll be out of pocket," or "I've got to go pay the bills." I told Carmen, "I'll be out of touch for a couple of weeks."

I couldn't tell her where I was going, so I used

this careful wording she would recognize to let her know I would be on a mission. I told her I would transfer some money to her account before I left so she wouldn't have any money worries while I was out. I told her I didn't know what the signal would be like where we were going, but I'd be in touch whenever I could, even just a word at a time, to say I'm okay.

I also gave a small suggestion: maybe do not watch the news for a few days.

You can only do so much to keep your family from worrying.

I got my affairs in order, and I left for Afghanistan.

This is where we jump back to the story at the very beginning of this book, when my teammates and I were on orders to capture or kill a high-value target in the hills of Afghanistan. We were several days into our mission, and the hunt was on. We knew they were watching us from the jagged mountaintops. Those guys were fast and quick, and they knew this land far better than we did. This high-value target would be difficult to track down, let alone pin down. If we had our chance to shoot, we should take it.

I had sights on the enemy, and I had my finger on the trigger. I knew this might be our only chance, but following the lieutenant's orders, I did not shoot. By the time we got close enough to capture them, they were gone. We had missed our chance.

We piled into our two trucks to travel again on that one road out of the village, to bring supplies to our teammates up in the mountains. The lieutenant drove the first vehicle, and I rode in the front passenger's seat. I put my lucky hat into the pocket of my cargo pants, and I watched the road ahead. Our Humvee drove through a shallow creek, and seconds later, the Humvee jolted and the ground shook beneath me.

There was a loud *boom*.

There was a flash of white.

There was intense heat on my left side.

There was a bitter taste of gunpowder.

And I knew.

Holy crap. We just got hit.

That is when those three images flashed before my eyes, visions I will never forget. My wife dressed in white. Our honeymoon in Greece. And my boy Guero, playing ball with me. I had a moment of clarity where my thoughts were clear as day, and I realized I had broken the promises I made to my family, to my son, and to my dad. The promises that had anchored my life came unmoored on the battlefield that day.

Something inside me shouted, Get out of this truck. *Now.*

I popped the door open and got out as fast as I could. I was already on fire from head to toe. I turned to run toward the water behind me—but the flames overtook me. I collapsed to the ground. I lay there, burning alive.

My lieutenant—the same one who had been so green and immature—came through for me that day. I heard his voice, shouting over the flames, "DT! DT! You are not going to die here!"

He leveraged himself underneath my shoulder to help me stand, and then he walked me to the water. It was more than heroic—it was a sacrifice. By standing so close to me, he caught on fire, too. We both jumped into the creek to extinguish the flames. I remember the sizzle sound as our bodies hit the water.

Lieutenant and I lay in the shallow water, breathing hard. The air was filled with smoke, my clothes had burned off, and my ears were ringing from the decibels of the explosion.

I said, "LT, man, this sucks."

I don't know if I was trying to calm the situation,

if I was trying to distract us from what was going on, or if it just made sense to me to say that. But those are the words that came to my mind.

Lieutenant looked at me. "Dude, are you trying to be funny right now?"

"No, man, I'm serious. I just got blown up. I was on fire, and now I'm lying here with you in a freezing cold creek in the middle of December in Afghanistan. This sucks."

"You could say that, DT. This definitely sucks."

Seconds later, I heard a call sign on the radio: "We need Gunslinger."

My radios—and backup radios—were destroyed in the explosion, but the radios in the second vehicle were intact. I heard my teammates call again: "We need Gunslinger."

At the same moment that the bomb hit our Humvee, the enemy hit our teammates with cross fire. They were under attack, and they were calling for help. They were calling for me.

One of my guys from the second vehicle ran into the creek to give me his radio. I remember the water splashing against his boots. He handed his radio to me, but I couldn't take it from him. That's when I discovered the first of my injuries—my hands hurt too much to grip the button. Instead of calling the orders into the radio, I told my teammate, "Repeat everything I say."

Lying in the water, my chest tighter with each breath, I called out the orders, and he repeated each one, word for word.

The medic arrived as we were calling orders. He came straight to me, but I sent him to check on Bailey, a gunner who had been in the vehicle with us, who had been blown out of the truck. I knew the Humvee had run over his legs. I said, "No, no—I'm fine. I have all my body parts. See if he's okay. Just, please, cut off my shorts."

We call them "ranger panties," those running shorts we wear underneath our pants. When my clothes burned off, the only piece that remained intact was the elastic band of those ranger panties. The rubber was burning into my waist. The medic cut the shorts off me, and I lay there—naked and burned—calling out orders for my teammate to repeat.

We got help for our brothers, and by the time the last phrase came out of my mouth, as my teammate repeated my words, something shifted within me. The panic was rising, the adrenaline was leaving the body, and my chest started to tighten. I couldn't take a full breath.

Listen, I'll never stand in front of anyone and tell them I was a brave guy who wasn't afraid. I'll never say I went out and did my mission without any fear. Yes, I completed my mission, but not without fear. When I couldn't take a full breath, I got scared.

I suddenly felt so sleepy. All I wanted in the world was to close my eyes, just for a few minutes. The aircraft were coming to help my teammates, and I knew I needed a medevac. I remember thinking, Where's the medevac? What's taking this helicopter so long to get here?

I thought maybe I could rest for a few minutes until the helicopter arrived. I told the medic, "Just let me close my eyes for a few minutes."

I didn't know how badly I was injured, and I didn't know that this sleepiness was actually my consciousness fading in and out. But my buddy knew. He knew if he let me fall asleep, I'd never wake up. He wouldn't let that happen.

You see, my buddy knew me. When we're out there in the field with our teammates, we have the gift of time together. Sometimes there are long stretches of hours between action. When we're sitting there with nothing going on, meal after meal and night after night, we start to tell our stories. We talk about the sports teams we've played on and cheered for. We learn the names of each other's hometowns, our pets, and our families—every woman and child waiting for each of us back home. We become brothers in the field, and we learn about each other.

Medic knew that I had lost my dad when I was young. He knew I had had to make my own way in the world, raising my brother and sisters. He knew how I loved my wife, and he knew the promises I had made to my son. He knew what I had said—that I'd never let my son grow up without a dad. He knew what was at stake.

He called out to me, "DT! Come on, man, fight! Don't fall asleep! Stay up!"

"I'm tired, man. Let me sleep. For just . . . a few minutes."

A few minutes would be forever. Medic knew that, even as I was fading.

He said, "Come on, DT. Fight for your son. Remember your promise. You said you'll never let him grow up without his dad, and you have to fight for him right now. Fight for your son, DT. Say it with me. Say you'll fight for your son."

"I . . . fight for . . . my son."

"That's right, DT. Say it again. Yell it. Yell it!" he said.

"I fight . . . for my son." My shout was a whisper, and it was all I had.

"Fight! Fight, buddy! Fight!"

"I fight . . . for my son . . ." I repeated his words, but my eyes were falling closed.

And then, I remember he said the weirdest thing. He said, "DT! You have to fight for your son so you can teach him to be a pimp!"

I remember thinking, even in that half-conscious moment, Did he just say that? A pimp? My son is three years old.

I knew my head was scrambled, I was confused, and I had probably heard him wrong. I opened my eyes.

Medic said, "Yeah, man! There you are, DT! Yes! Fight for your son so you can teach him how to be a pimp!"

So, there I was, butt naked in Afghanistan, lying in a river, and screaming at the top of my charred lungs, "I got to fight for my son! Teach him to be a pimp!"

It makes me laugh to write about it now, and it is a funny memory. A funny memory in a terrible moment.

But, you know what? Medic did what he had to

do. He reached into his catalog of facts he had learned about me while we served together, and he made me say some weird shit just to keep me awake. He did what he had to do to keep me alive.

You do whatever you can to keep your teammate going.

He kept me awake until the helicopter came. I remember when the rescuers arrived on the scene, camouflaged figures kneeling around me, preparing to carry me to the helicopter.

Before they could lift me, I sat up, suddenly alert: "Oh, hell no. You're not carrying me. I walked into this fight. I'm going to walk out."

I got up. I hobbled my naked butt to the chopper. It was slow, but I got myself into the helicopter.

From there, the memories play like clips from a movie.

I remember lying back on the gurney, resting my head, thinking, Thank God, I can rest now.

I remember the flight landing at our forward operating base.

I remember going to our field hospital.

I remember seeing the faces of some of my teammates.

I remember them asking how I was doing, and I remember struggling to answer.

I remember the doctor cutting off my watch.

He said, "You're going to be okay."

Finally, I closed my eyes. For a long, long time.

CHAPTER 9

When you listen to the story of someone injured in war, the storytellers—whether they are reporters, politicians, people raising funds, or the leaders of an organization—often like to focus on the service member who got hit. Sometimes they forget to mention that our families are in it with us. There are parts of the story that are hardest for them.

Our families see us at our very worst. They're the ones who go behind the closed doors, who have the hard conversations and make the most important decisions along with the doctors—to amputate or not, to fight infection, or fight death. The military families are the ones making the decisions, and they are the heroes.

My memories pressed pause on December 4, and

my wife's chapters of our story began that day, when she received the news. Everything I know about that day is from her perspective.

She didn't have a computer in her home in Mexico, so she went to a nearby internet café every day, paying by the half hour to check her email. It had been a few days since she had heard from me, and the last email said only two words: "I'm okay." Literally, just those couple of words were all I had time to say. She was holding on to those words, waiting for some more.

As promised, I had wired money to her every few weeks, and she stretched the cash as far as she could, never splurging on frivolous spending. She knew I would be out of pocket for a while this time, so if she needed any money, she would borrow from her mother, and we would pay her back when I was able to get online again. I didn't think it would be a problem. We thought it would be only a few days.

Carmen remembers the last time we spoke on the telephone before that mission downrange, when I had told her I would be out for two weeks, maybe more. I knew it was a dangerous mission, but I didn't tell my wife. No reason to make her worry when there was nothing she could do about it. She remembers it was harder to say goodbye, harder to hang up the phone, as if we both knew something was going to happen. After Guero's emotional goodbye at the airport, and then that last phone call with me, she felt like a dark cloud was looming. Something bad was going to happen.

Carmen says that when she thinks about that last night before everything changed, that's what she remembers. The bad feeling.

Even though I had discouraged her from watching the news, still she watched every single night. When a husband is out of touch on a mission, the wife will lean on any updates she can get.

When she went to bed around ten thirty or eleven on December 4, 2005, it was already December 5 in Afghanistan. She couldn't sleep at all that night, tossing and turning in her bed. She was so tired, but her brain would not stop spinning with the notion that something was wrong.

She walked through the kitchen in the early hours, getting a glass from the cupboard and filling it with water, all without turning on the light. My wife remembers standing at the sink, praying for her husband. Carmen is a very devoted Catholic woman, and she learned long ago: when there's nothing you can do, praying is the best thing you can do.

Still, she could not get to sleep. She watched the sun come up outside her window that Sunday morning. December 5 was her birthday.

Guero and Carmen had a nice, easy morning with her parents, and she planned to go to the noon service at church. She loved that later mass, so she could spend the afternoon on the little plaza, enjoying a late lunch with her friends.

My boy and my wife walked several blocks to church each week, and they always sat in the pew with many of her friends and their children. They sang together, they listened to the priest, and they took the bread and the wine of the Eucharist communion.

On that morning, as the church service ended and the priest began to pray, Carmen felt a commotion behind her. She looked over her shoulder to the entrance of the church, and she was surprised to see her sister standing there. She was waving at Carmen to come to her.

Carmen thought, *What in the world? What is she doing, waving to me during church?* She knew she'd see her in a few minutes, after the service, so she looked away and bowed her head to pray. But when the priest finished praying, her sister was there at her side.

She said, "Carmen, *tienes que venir conmigo.* You

must come with me. Junior's friend called. He is going to call again in five minutes."

Carmen said her stomach dropped that very instant, and she knew. She grabbed her sister's arm, and she said, "Something happened to Junior."

"No." Her sister shook her head. "I think it's just something about paperwork."

"No." Carmen gripped her arm tighter. She said, "I know it. Something happened to him."

She recalled my parting words, preparing her for a moment like this. *If one of my friends calls you, that means something bad has happened, and I may be in the hospital. But if my commander calls, and a priest shows up at your door, then that means goodbye. It means I'm no longer with you guys, and I'm with my parents.*

She had braced herself for this phone call. But her heart could never have been ready.

Carmen turned to the friend on the other side of her, Liliana. She asked Liliana to keep Guero while she ran home, to please bring him home with her.

It was about a ten-minute walk from the church to Carmen's family's home, and she had to cross a high traffic area, an interstate with a lot of semitrucks. There was always a long line of traffic on that road, and Carmen and Guero always had to wait to cross the street. But that morning, it was as if the roads opened for Carmen. There was no waiting, only open road. She said she had never run so fast in her life. She felt like she was floating. She couldn't feel her steps on the sidewalk.

She got to the house in just five minutes, and the telephone rang again. Her heart was racing, and she was out of breath from running. She picked up the phone to hear the voice of my friend who spoke Spanish.

"Carmen, how are you doing? *Cómo estás?*"

"*Estoy bien,* I'm good, but I know something happened to Junior. *Dígame que está pasando.* Please, tell me."

"*No, no te preocupes,* don't worry. Nothing bad has happened to him. He will be fine."

"*No, por favor. No mientas,*" she said. "Please. Do not lie to me."

She heard him draw a breath, and then he said, "Carmen, I need to tell you, there has been an accident. *Hubo un accidente, y él está en el hospital.* He has been injured, yes, and he is in a hospital. But I want you to know that he's awake, and we believe he's going to be just fine. *Él estará bien.*"

The commander was on the phone as well, but he

could only speak English. He spoke next: "He'll be fine, Carmen, don't worry. We will take care of everything, and we will take care of you and your son. You are safe with us. We are here for you. I want you to hear me: DT is going to be fine."

They lied to my wife.

I understand that they needed to keep her calm, so they did not tell her the whole truth. But on the day that I was injured, Carmen didn't get all the information. Even if she could have understood English better back then, Carmen said his words sounded garbled and confusing. There was so much to take in.

During that first phone call, they didn't tell her to come. Not yet. They told her that I was injured, they told her that I would be fine. They gave Carmen their phone numbers so she could contact them, and the call was finished. Carmen says the conversations after that first one all live together in her mind like random puzzle pieces. She didn't even really know what to tell everyone. She couldn't answer anyone's questions because she didn't have any answers.

She told her parents what she knew, and then she called to deliver the news to my sisters. Together, they started a chain of phone calls throughout the whole family, all the aunts, uncles, and cousins throughout

Chicago and Mexico. I have always been really close to my family down in Mexico, and all of my aunts and uncles saw me as a child of their own. The news was devasting to everyone, and even more so for them.

My uncle had been out running errands that morning, and he stopped by the house as soon as he got the news. He was crying as he said, "Oh, Carmen, I know what happened to Junior. I just want to let you know that we love you. We love you, and we love Junior. If you need anything at all, we've got you."

Carmen's sister called to wish her a happy birthday, and her voice was so happy on the phone until Carmen told her what had happened. If I could change just one thing about this event in our lives, I'd pick a different day for the explosion. I hate that all of this happened at all, but I especially hate that it happened on Carmen's birthday.

All through the day, the commander called every hour with updates. Every phone call brought more bad news, and they broke it to her slowly.

"We need to tell you that he has been burned. But it's only a little bit, Carmen. Only ten percent of his body."

Next phone call: "Carmen, he's burned on twenty-five percent of his body."

Next phone call: "Carmen, he's burned on fifty percent of his body."

They said I had been injured under fire, but that I was alive. They didn't explain anything about a bomb or an explosion. They told her I was fine, but she learned later that they weren't sure if I would make it through the day.

Late that evening, they told Carmen that I had been flown from Afghanistan to Germany, and that I was scheduled to arrive in San Antonio, Texas, by December 7. My commander said to my wife, "We need you to meet us there, Carmen."

"But, how?" she asked. "I don't have a visa. I don't have money. I don't have anything."

She thought about the money I had wired to her the week before. She realized that deposit would never be enough for the journey ahead.

The commander said, "Don't worry—I will do whatever needs to be done, Carmen. First thing tomorrow morning, you call the embassy near Juárez where you got your visa, and ask them to give you a permit. They have all your visa information on file, and I will confirm that they have all the information they need from us. They will be able to take care of it and get you across the border."

He gave Carmen the address and phone number to contact the embassy, since she didn't have a computer in her home.

The next day, she would travel to the United States with Guero. For the second night in a row, she lay awake, praying to God to spare my life. This time, she knew why she had been sleepless the night before.

Her soul had known.

CHAPTER 10

The next morning, Carmen was out of bed before sunrise, watching the clock for the hour when the office would open. She spoke with a woman at the embassy in Juárez. She gave her name, information, visa number, when it was issued, when it expired—every bit of data that she had. Carmen told them all of my information—she explained everything. She said, "Please, help me. I need to be there on December 7."

She could hear the woman's fingers clicking on a keyboard, and then the woman said, "I do not have any of your information here."

"What? How is that possible? I was in your office two years ago. I was there, right where you are. You guys took my fingerprints and conducted interviews to learn everything about me. Now you say you have nothing?"

"Ma'am, there seems to be nothing I can do for you," she said.

"Nothing? No, how can this be? My husband is a service member who has been injured in a war for his country, and I don't even know how bad his injuries are. They won't tell me that, and now you say you can do nothing to help me? I need to get there. Please," she cried, "I am begging you."

With a clipped tone, the woman said, "Ma'am, I suggest you try to cross the border and see what happens."

Carmen hung up the phone, her hands shaking in fury. After all the things we had given to the United States already, all these years of sacrifices, all these years of being apart. And now I was fighting for my life, and they told my wife to "try to cross the border and see what happens."

I wish I had the name of the person at the embassy who took her phone call that day.

As it turns out, my commander thought the very same thing.

He said, "Are you serious, Carmen? That's what she said to you? Give me her information and I will make a call."

That's when Carmen realized she didn't get the employee's information. She may have said her name at some point, introducing herself on the call, but Carmen didn't write it down. Who could blame her? She had so many things to keep track of, and her brain was scrambled. Nothing would stick. Everything would slip away.

The commander said, "Carmen, this is not your fault. She said the very worst thing that an officer can say to a service member—or in this case, to his wife. I'll make a phone call directly to the embassy in Guadalajara. I promise you this: you'll be in the United States on December 7. We need you to get here."

Before the end of the day, he called her back with a plan.

He said, "I've spoken with the embassy, and they are waiting for your call. They will see you in the morning to complete the paperwork, but you must have an appointment, so call them right now."

She followed his directions exactly. She called the embassy, and they required her to pay one hundred dollars just to get the appointment. Carmen remembers thinking, *Whatever. Yes. Anything. Take the money I have. Just get me to my husband.*

Once she made the payment, they talked her

through the same labyrinth she had navigated four times before, when she was trying to get her tourist visa. She already knew how it worked: you make a call, get an appointment, come before six in the morning, stand in the line that stretches three or four blocks around the building. In her experience, that line was at least a two-hour wait. When it is your turn next in line, you have an interview with an officer. Carmen had done this many times before, and while I was injured and fighting for my life, she would do it again.

Meanwhile, by contrast, inside the United States my family received a private military escort from the San Antonio airport to the hospital. I am thankful they took care of my sisters and my brother, but I wish they had done the same for my wife. Just south of them, she received no special treatment for being a military wife—even to an injured airman.

Carmen went to the embassy on her own, waited in the two-hour line, and finally had an interview with an officer. They confirmed that I am a citizen of the United States, working in the military. Guero, born to a US citizen, is also a citizen. Everything on their digital file matched everything in her hand. Even still, they authorized only a three-month visa for Carmen to visit the States.

They gave her a printed visa that looked like an American passport, complete with a temporary Social Security number. Below her photo, where most passports have a barcode for scanning, they printed a note in large letters:

> Wife of Airman, Sgt. Israel Del Toro.
> Airman injured in battle, in critical condition.
> Deliver promptly to Brooke Army Medical Center.

With the temporary visa in Carmen's hand, the commander booked her flight. The earliest she could arrive was at 7 P.M. on December 7, approximately twelve hours after I had been transported from Germany. The commander had hoped to get her there before I arrived, but it was impossible. She would get there as fast as she could.

Once her ticket was booked, the commander told Carmen the last hidden fact of my situation, the piece they had kept secret. He said, "Carmen, there's one more thing I need to tell you before you see your husband tomorrow. DT is in a coma."

One more blow of information, the truth of my condition. They were worried she wouldn't be able to handle the magnitude of this much information as she navigated the labyrinth of the embassy and the border, but now she knew what they knew.

My wife said evenly, "Thank you for telling me, Commander." She hung up the phone.

Nobody knew yet that I would be in this coma for four months. The temporary visa she had just received wouldn't guarantee her enough time to stay by my side until I woke up. Nobody could know how long this would last, or if I would make it at all.

When Carmen tells the story of that morning, she recalls standing over the bathroom sink, looking at her reflection in the mirror. She saw how her face had changed in the last two days. Her eyes were puffy and swollen, the skin around them aged and tired. It had been the hardest couple of days of her life, and she was headed into a battle she couldn't wrap her mind around.

I always smile when I hear this next part of the story—how Carmen gasped out loud when she saw her hair in the mirror. It was dyed a crazy shade of blond, an experiment by her sister. Her sister is a hairdresser, and they were always doing bizarre stuff with Carmen's hair. It was harmless at the time, but now it felt like a huge mistake. She couldn't even look at herself in the mirror with this ridiculous hair, and she definitely didn't want me to see her like this. There was a chance I might not recognize my wife when I saw her, but Carmen couldn't let some strange hair dye be the reason I didn't know who she was.

She called her sister and said, "*Mi pelo.* My hair. You have to fix this."

Everything had fallen apart. The only thing she could fix was her hair color. On the worst days of your life, you fix what you can fix. Even if the only solution comes in hair dye from a box.

Carmen and Guero had been in Mexico for six months since that disastrous trip from Italy with a detour through Amsterdam. She packed only what they needed to come to the United States, which came down to three suitcases to carry everything for the two of them. Of course, she had no idea how long she would stay.

The nearest airport was more than two hours away from her home, so they had to leave three hours ahead of time. Carmen said goodbye to our families, all of them crying, praying, and wishing the best. Her mother gave her one hundred dollars cash to carry in her wallet, and she boarded a plane with baby Guero, straight from Guadalajara to San Antonio. By the time they arrived, it was after seven that night.

Finally inside the San Antonio airport, she gave her visa to the officer at customs. The paperwork was as comprehensive as an American passport, complete with those additional details about me being in critical care. This should have made it easier for her to get through, and that note from the commander should have been enough to get her directly through customs. But the officer barely glanced up from his computer.

He looked at the visa, typed a bit on his keyboard, and then he said, "Ma'am, why do you have three suitcases? Why would you need so much luggage for a short stay?"

"Sir, I have a letter for you from my husband's commander, explaining why I am traveling and why I may need to stay a long time."

She pulled it out of her bag and placed it on the counter.

He glanced at the typed page, and he returned to typing on his computer. Then he said, "You need to follow me. Come this way."

Carmen and Guero followed him down a hallway, into a tiny room with a desk and a chair. In that closed space, with our little boy, her suitcases, and all her paperwork, the officer and his partner peppered my wife with questions.

"Why are you coming to the States? What is the story with this visa? If you're not a student or a tourist, then why are you coming? If your husband is a citizen and your son is a citizen, why aren't you a citizen?"

None of her words were enough. She had to answer the same questions and explain herself over and over and over and over again. The interrogation lasted for over an hour.

She finally said, "If you don't believe me, please. Let's stop doing this. My husband's commander is waiting for me inside this airport, and you can speak to him yourself. My husband's family have been escorted to the hospital, and they are all waiting for me there. If you do not believe me, please, just go and talk to any of them—the commander or my family. If you don't want me to be in your country, that is okay with me. But please, let's stop this."

She broke down and wept. It was all too much. She thought, *Isn't it enough that my husband is injured? Isn't it enough that he is lying unconscious in a hospital bed? Isn't it enough that he is dying? But now I cannot get to him? What else could go wrong? What else would have to happen to end this nightmare?*

She cried to the officers, "Can't someone please help me? I have shown you everything that I can to assure you that I am not here illegally. I have to be a resident for three years before I can become a citizen, but my husband has been deployed all over the world, and that is why I am not a citizen yet. I cannot make you believe me. But please, can you help me? At least let someone know. Let someone know that I tried to get to my husband. Please."

I wish I could have protected my wife and my son from this whole interaction. Guero was getting restless. Imagine any little boy in that situation, locked in a room with a desk and a chair, while his mother is questioned. It was stressful for him, too. I think that may be what caused the officers to finally consider letting them through: my son was going crazy in this little room.

Finally, they concluded that she wasn't lying, and they allowed Guero and Carmen to pass through customs. By the time they got to the hospital, it was almost nine thirty at night.

My aunt Griselda is from my hometown, and we call her Tía Gris. She has been like a second mother to me. She loves me like her own son. She met Carmen at the hotel next to the hospital, and Carmen said she wept when she saw Tía Gris's face—finally someone who knew her and loved her. They cried together, heartbroken.

Tía Gris had already been to the hospital, and she had seen the condition that I was in. She tried to prepare Carmen for the shock of my appearance.

She took Carmen's face in her hands, and she said, "Carmen, Junior is not the same as he used to be. *Estás preparada.* Be ready for what you are going to see. Be prepared, and be strong. He is not the same. You need to be ready."

How can one ever be ready?

Still, she nodded. Carmen said, "I just want to know the truth. And I want to see my husband. Just let me see him."

Tía Gris walked with Guero and Carmen from the hotel to the hospital. It was only across the street, but it was a long journey. Carmen had never been inside a hospital's burn unit before, and this was a whole new world. So many long halls, bright fluorescent lights, and dozens of elevators. Such a strange energy there; the nurses with the clipboards, the low tones of people talking to one another, a TV playing in a waiting room. And the smells were so conflicting, bleach and urine and chicken soup.

Gris stayed with Guero in a waiting room while Carmen waited to enter the locked doors of the intensive care unit, an environment all its own. They opened the sliding door for Carmen to enter the burn unit, and the smell was even stronger in there. So many chemicals, medicines, liquids, and all the sterile dressings covering burned flesh.

She met the nurses assigned to care for me. They gave her a gown, a face mask, a hair covering, and gloves. She was covered from head to toe.

The doctor spoke very seriously to my wife. He said, "Mrs. Del Toro, you must understand that you absolutely cannot cry inside this room. We cannot risk any bacteria from your tears, but also, we don't want him to hear you cry. We're not sure what he can understand, but research shows that patients can hear you. Patients can feel your presence while they are in a coma, so this must be a positive experience. It's best for both of you if this is not an emotional experience, so you cannot go in until you are sure you will not cry."

She gathered herself, taking slow breaths to contain her emotions. When she said okay, they opened the sliding door to my room, where she was finally allowed to see me for the first time since my injury.

Carmen recalls that I was wrapped in white bandages from head to toe. She said my face was swollen like a basketball, but it was the deepest red of a radish. There were so many tubes and wires, so many machines beeping around me and breathing for me. My arms were stretched out at my sides like the letter *T,* and the only parts exposed were the tips of my fingers and the tips of my toes. They were black as charcoal, no flesh. Just black bone.

She said she so badly wanted to stay by my side, but she was allowed only a few minutes. She could not touch me, and she could not cry, but she had seen me with her own eyes. Now she knew what they had not told her. Her husband was dying.

CHAPTER 11

My wife and my son made their temporary home in a tiny hotel room on Fort Sam, with a microwave, a bed, and a few belongings. Every day, Carmen walked to Brooke Army Medical Center (BAMC), leaving Guero with my brother or Tía Gris, while they were still in San Antonio with us. Guero was never allowed inside the ICU, and Carmen could not come until noon because the nurses needed hours every morning to change my dressings and tend to my wounds.

So, she spent every morning in conversations with the government, trying to acquire and provide information, asking permission to stay inside the United States. I was in critical condition, living just one moment at a time. If at any time I died, she knew she would likely be deported the same day. Certainly, they would send her away without all her belongings, and perhaps without our son.

Carmen became fastidious about every detail. She bought a spiral notebook, and she started an expense report to keep track of every penny, coming or going. She wrote down every exchange: when she received a gift card from a friend, when the Air Force gave her per diem money for food, when my buddy Jeff Mariano came into town and gave her some cash to carry her through. She recorded every detail in her notebook.

Remember, Carmen had come from Mexico with the one hundred dollars in her pocket, the cash her mother had loaned to her. She didn't have a credit card, ATM card, checks, or any additional cash. She could not access any of our accounts without an ID, which she couldn't access without a permanent Social Security number, which she couldn't access without a residency card. One problem kept leading right back to the problem before it.

Aside from the administrative tasks of paperwork, passports, and finances, Carmen had to manage the ever growing tension from my sisters and brother. I can only imagine how terrified my siblings were, after they had experienced so much loss. I had become like a father to them, the patriarch of our family. How helpless they must have felt, to relinquish any sense of control. Trauma affects people differently, and people can act scary when they're scared. They were not kind, but they were not themselves.

The tension was not over the decisions Carmen was making, but it bothered them that the doctors were constantly turning to her for authorizations. That's how it is in crisis situations: the hospital needs a point person to make decisions, and the primary contact is the spouse first. The hospital had rightfully put my wife in charge.

But my family argued with Carmen over every decision she had to make. They didn't want her to authorize any decisions about my medical care or my belongings. There were so many things to consider—my car, my motorcycle, our money—all of it. They didn't believe Carmen could process the information or make the necessary decisions, so they said such mean things to my wife. Their words still hurt my soul, to consider how they had hurt her back then.

They claimed that she was incompetent because she came from a small town, she didn't have a driver's license, and she didn't finish high school. No, she didn't speak English, and she didn't drive. Those are true statements. No, she did not have a high school diploma, but she is not without an education. She had finished ninth grade and then started at a technical school to become a bookkeeper. Carmen knew how to plan out a budget, write a check, and make financial decisions. But they questioned everything she did.

There was even a time when my brother asked Carmen to account for all the money she had, any money she received from the Air Force, any gift cards from friends. He wanted an expense report of everything she had received and everything she had spent. Why? Because the extended family suspected she was sending it all to her family in Mexico.

She showed them her notebook, every transaction, every penny. Carmen hadn't even paid her mother back the small loan, let alone sent them anything extra. She was honest about what she had, and she wasn't stealing from anyone.

As my wife, Carmen was the point person, and she was capable of doing everything that needed to be done. My brother and sisters all had finished high school, but they were the ones who looked uneducated. My siblings had each other to lean on, but Carmen was there alone. She had only our son, and this heavy pressure was all on her. Yes, I can imagine how hard it was for my brother and sisters, but their criticism made everything harder for my wife. Imagine being in her place—navigating another culture, language, and transportation system with the stress that she was under.

When I think about that, damn. I want to give the woman a diamond ring and a superhero cape.

A few days after my injury, my family traveled back home to Chicago for a week or so, and then they came back to San Antonio for Christmas. Carmen's brother and sister each flew in for a few days at a time, but they couldn't stay long either. Everyone had their lives to return to, and that meant Carmen didn't have anyone to help her with our son.

There was no daycare, because it cost money she did not have. Even as the military began to help me with costs, the childcare on the military base would have to come from her pocket, and she just did not have it. For that entire first month, when no one else was there, Guero was with Carmen all the time.

When she had permission to visit me, for ten or fifteen minutes a day, she could not take Guero into the room. She had to ask strangers to keep an eye on him while she scrubbed and gowned herself to visit my bedside for a few minutes at a time. Carmen still cries when she talks about the hardest part of all, dividing her time between her husband and her child, constantly wondering if the other was safe when she was away from him.

In the first few weeks, she met many families in the waiting rooms of the intensive care burn unit of BAMC. The burn center had cared for thousands of wounded warriors who had been evacuated from Iraq and Afghanistan, along with thousands of civilian patients from the south central Texas region. She met many people who were in the same boat. Some of them were military spouses or parents, and others were civilian families whose loved ones had been in an explosion or a car accident.

That is when Guero and Carmen met Vickie. She was a teenage girl, not yet eighteen, and she spent her days alongside them because her grandmother had been injured and burned when her oven exploded. She befriended Guero in the waiting room, and she stayed with him for the twenty minutes each day so Carmen could visit me. Guero called her Tía Mickey, like Aunt Mickey Mouse. She was his favorite babysitter, giving him anything he wanted.

In his three years of life, Guero had never had a taste of soda, but Vickie bought it for him from the hospital vending machine. Carmen didn't even mind. You fix what you can fix, and you can't worry about everything. She let Vickie spoil our little boy with a can of orange soda now and then, as she carried him on her hip around the hospital. I smile at the thought of it. Vickie was our angel.

There were two men who stayed with my wife all the time, never leaving her side: Chief Humphries and Chief Gonzalez. Chief Humphries was the family liaison, communicating all of the information between the Air Force, the hospital, and Carmen. He was there waiting when she arrived on day one, and he watched over my family every single day. Any time the hospital had information, they called Chief Humphries at the same time that they called Carmen. If they called her, they called him also, so neither one was ever left out.

Chief Gonzalez was a master sergeant in the Air Force at the time, and he served as the translator who could make sure Carmen understood every word of information from every doctor and nurse. There were some doctors, nurses, and hospital volunteers who spoke Spanish, and Chief Gonzalez worked together with them to relay every word to her.

When I got hurt, my teammates came from all parts of the world to be by my side. My best buddy, Jeff Mariano—remember the scrub in the passenger's seat of my Jeep in tech school?—came every month. Dude was planning his wedding, and he put the whole thing on hold to make sure my wife and my son had the support they needed. My very best guys were there nonstop.

When my wife had given me the ultimatum before I left for Afghanistan, she couldn't understand why I wanted to stay in. But when I got hurt, she saw the brotherhood of mine. For the first time, she understood why I wanted to serve alongside these good men. The military is a battle, and they fought right beside her. They had her back, and they had our son's back.

When Carmen's brother visited for a few days, he saw what Carmen's mornings looked like. Chief Humphries and Chief Gonzalez drove her to appointments all over the city before her daily visit in the burn unit. She was always traveling back and forth—back to the hotel to make a phone call, back to the hospital to meet with the doctor as he did rounds, off to an Air Force base to complete paperwork, off to a bank or Social Security office. Wherever she was, someone needed her to be somewhere else.

Her brother said to her, "You need a cell phone, sister. Everybody is trying to reach you, and we need to know where you are."

Carmen was in absolute agreement; a cell phone would have made a few things much easier. But again, she couldn't get a cell phone because she didn't have a bank account in the United States, and she didn't have a bank account because she couldn't get a Social Security card. After much runaround, the team at the T-Mobile store next to the hospital were the ones to finally help her get a cell phone. They required a five-hundred-dollar deposit of cash, and her brother footed the bill. At last, she had a cell phone.

Carmen says she will always remember the first time it rang in the middle of the night, around two in the morning. I had been in the hospital for nearly a month—more than twenty days. (She was counting every day.) She kept her phone plugged into the wall throughout the night, charging even while she kept it in the bed with her. She answered on the first ring.

The nurse said, "Mrs. Del Toro, we need you to get to the hospital immediately. There is something going on with Israel, and he is not responding. We need you here now. Your husband may not survive the night."

Carmen's sister was staying with her, and she leapt out of bed when Carmen woke her. They threw on their jackets over their pajamas, and they slipped into their shoes. Carmen wrapped Guero in a blanket, and together they ran the four blocks to the hospital, handing him back and forth when one of them got tired.

When they arrived in the ICU, Chief Humphries was already there. The hospital had called him at the same time. My sister-in-law and Guero fell into a chair, and Carmen approached the nurse's station, holding her chest and catching her breath.

The nurse seemed surprised to see her. "Mrs. Del Toro, how did you get here so fast? Did you get a ride?"

"I ran," she said, her breath uneven.

"Oh, my God," she said, "you are so fast. I thought maybe you took the shuttle bus."

Carmen thought, *Wait, there's a shuttle? Nobody told me about a shuttle bus.*

For weeks, she had been walking—sometimes running—between the hotel and the hospital. She had come at all hours of the day and night, carrying our baby on her hip every time. All along, there had been another way, a faster way, but nobody had told her there was any other choice. The information would have been helpful.

The nurse said, "Well, now that you're here, we can tell you what is going on with Israel."

She came from behind the counter to sit with Carmen, to explain that I was having trouble breathing. Even with the respirator, I could not take in enough oxygen. She explained that a healthy person has baby hairs inside their bronchial airways, and they act as a filter for the air we breathe. I had inhalation burns inside my body, and those hairs were burned away. With nothing filtering my air, there was now a buildup of mucus blocking my airway. They needed to open the side of my rib cage to get to my lungs so they could suck out that thick mucus.

She said, "We are going to try to open his airway, but I need to tell you that it doesn't look good. I recommend you begin calling his family to let them know, and in the meantime, try to relax in the waiting room."

Relax? Who could relax in a moment like this?

Chief Humphries, Carmen's sister, Guero, and

Carmen waited through the night in that waiting room. They made phone calls, and they watched the clock. But Carmen said she was sure of one thing: none of them relaxed.

Hours later, at seven thirty in the morning, the doctor came into the waiting room. He told her that the procedure had gone as well as they had hoped, and I was stable. I would live to see another day.

He was carrying a specimen cup, the same kind you pee in at the doctor's office. He held the cup for Carmen to see inside, and she saw a black-brown stone the length and width of her thumb. That's what they had removed from my lungs. It rattled like a rock in the cup. My lungs were filled with stones like this one, made of blood, mucus, smoke, and charcoal dust.

For weeks, Carmen had been unable to eat. Just a little bite of food made her feel like she might throw up, and the only foods that she could stand were small bites of yogurt, Jell-O, and the occasional sandwich. All of her stress was in her stomach.

She has told me that she will never forget that New Year's Eve, her first meal in a month. Her sister bought a small roasted chicken and a bottle of sparkling cider from the grocery store. In their little hotel room, they raised their Solo cups together, a toast to the new year.

Feliz año nuevo. Buena suerte.

Happy New Year. Please, may there be better things to come.

In the new year, and in addition to navigating these sleepless nights, hard decisions, family conflicts, and a language barrier, Carmen set a new goal. It was time to learn to drive a car.

There was a lot of traveling involved for the many appointments, from Fort Sam, to Randolph Air Force Base, to Lackland Air Force Base. Each one of these bases is clustered around San Antonio and Corpus Christi, and they are pretty far from one another. It wasn't efficient for military personnel to drive her all over Texas day after day, and she needed some independence.

Also, nobody knew if I would wake up, but there was reason to plan with optimism. Even if everything went as well as it could possibly go, there would be a long process for rehabilitation and recovery, and I certainly wouldn't be able to drive a car. That would be her job.

The Air Force paid for her driving lessons, and my teammates arranged for my car to be delivered to Carmen in Texas. It was a bumpy road to get her license—

both literally and figuratively—but she did it. Add it to the list of her accomplishments.

And there's one more list to speak of: Carmen's list of the angels who sheltered our lives in those months. Each one saved her life and mine.

There was Chief Humphries, who was with Carmen every day.

There was Chief Gonzalez, who made sure she could understand every word.

There was Vickie, whom Guero still calls Tía Mickey.

There was Staff Sergeant Olga Hudson, who took Carmen to her driving test, to her English classes, and to a hundred appointments all over town.

There was the teller at the bank, who listened to her story and helped her open a bank account, even without a Social Security card.

There was Father Jimmy, the hospital chaplain who introduced Carmen to an immigration officer. With this missing piece, Carmen was able to complete the necessary paperwork so she could stay by my side.

There was the immigration officer who accelerated the process, granting Carmen's residency immediately, in a process that only took a few minutes. She didn't even have to make an appointment.

And by the way, remember that temporary visa they gave her at the customs office? That Social Security number they gave her on that first day in the States—that is the same number she has now. It was good from that moment forward, and she could have used it all along to get the support she needed. But for reasons I'll never understand, people kept putting barriers in front of her, making everything harder than it needed to be.

If you have the chance to make something easier for someone, don't make it harder.

Be someone's angel.

CHAPTER 12

The nights were the hardest on both of us. Between January and March, there were three times the hospital called Carmen while she slept. Each time, they didn't know if I would survive. Each time, she stayed awake in the waiting room, rocking our child, wondering if his father would live until morning.

Carmen had always loved that eighties song by Richard Marx, "Right Here Waiting." Even before she could speak English, she had loved the melody of that song. But now that she knew the words and what they meant, that love song carried her heart through those long nights as she prayed I would survive.

Wherever you go, whatever you do, I will be right here waiting for you.

In January, I began to experience further complications. The infection was getting worse, and the doctors became deeply concerned about my fingers. They had run tests to see if there was any live tissue in the fingers, but they found none. The fingers were dead, crisp charcoal stones on each hand.

The specialists explained the complications to Carmen. If they kept my fingers, the risk of infection was severe, and I could lose more of my body—perhaps an arm. Carmen needed to decide if they should amputate my fingers, in hopes of saving the rest of my body. What a difficult decision to make for someone else, and she couldn't even ask me what I wanted.

She told me how she agonized over the choice. She thought of how I loved to ride my motorcycle, how without fingers, my hands would never wrap around the handlebars again. She knew how I loved to play baseball, and how my hand would never slide into a glove again. She thought of my independence, of the countless reasons a person needs their fingers.

But then she thought of the alternative. If I didn't

lose my fingers, I might lose my arms. Maybe even my life.

She could ask for advice, she could seek counsel, and she could research the options. But ultimately, the hospital required my wife's signature on the consent form. She only wanted me to be well, to recover, and to live, so she made the hard decision. She signed off on the procedure, knowing I would never have my hands whole again.

Then there were the decisions to make with experimental medications. The doctors had tried them on another burn victim, and he had experienced favorable results. That man was awake, standing up, and walking. So, the question was on the table: Should they try these experimental drugs on me, even though I was not yet awake from the coma? There were no guarantees—it could go very well, or there could be long-term side effects that they didn't know about yet. The chances were fifty-fifty, and there wasn't much time to decide. She had to let them know by Monday.

To help her make the decision, they gave Carmen a book written in two languages—the first half in English, the second half in Spanish. She had one weekend to read, research, learn, and decide which decision would be best for the fate of our family. Again, my siblings were conflicted about the best decision. My sisters didn't want them to give me any experimental drugs, but how could Carmen just say no without learning everything she could? What if this was the answer? What if this could save my life?

All any of us can ever do is our best. She made each decision with the information she had, according to what seemed best for our family and for me. After all, every decision affected all of us.

On Monday morning, Carmen told the doctors that she had decided not to consent to the experimental medications. She made her choice because she knew it was the right one, not because anyone—not even my family—told her what to do, gave her advice, or weighed into her process. She informed the doctors of her decision, and she could only hope and pray it was the right choice, that her heart was guiding us in the right way.

Everything was hard for a long time. Yes, it was hard for me, but I do believe those weeks were harder for Carmen. I was fighting for my life, but she was fighting for my life, too. And she didn't have the benefit of sleeping through it. I will never be able to fully understand the weight of those long days; so many decisions, how she felt crushed beneath each one.

While she could talk to me, so often she just wanted to hear me talk back to her. She longed to know for sure what I wanted and needed, to ask me what she should do next. It was the darkest time of her life, as she fought for me every single day that I was asleep.

Each day that Carmen sat by my bedside, she talked to me. Nobody knew for sure if I could hear her, but they believed I could sense her presence, feel her energy, and maybe some part of my sleeping brain could recognize her voice. She was willing to try anything, and she spoke to me every single day.

"Remember your promise, Junior," she said. "You promised your dad you would take care of your family. You promised me you would come home. And you promised Guero you would never leave him. Keep your promise, Junior. Please, please keep your promise."

I was hooked up to so many monitors that kept constant watch over any shift in activity while I was in the coma. The doctors and nurses in the intensive care unit were monitoring always, careful never to miss any critical information.

One day, as Carmen talked to me about Guero, the heart monitors showed a difference in my heart rate. At the mention of our son, something in me responded, and my heart rate went crazy. She called out to Chief Gonzalez to watch the monitors with her as she tested her experiment again. This time, as she talked more about our son, they both saw that I began to cry. Nothing like this had happened ever before.

When the nurse came by to ask how the visit had gone, she asked Carmen if she had seen anything new. The answer had always been "no," but this time it was different. Carmen told her, "Do you know what? I talked about our son today, and Junior cried. *He cried.*"

This was a wonderful indication of what might be happening in the cerebral cortex of my brain.

Then Carmen had an idea—an experimental therapy all her own. If my heart rate responded hearing *about* him, what might happen if I could hear his *actual* voice?

She was not allowed to bring anything into the room because of the constant risk of infection, but she wondered if there could be an exception. If they used antiseptic sprays and sanitizers to be sure there was no contamination, could she purchase a tape recorder and capture Guero's voice for me?

The doctors said yes, it was worth the risk to give it a try.

Carmen bought a small recorder, and as she played with Guero in their tiny hotel room, she made an audio recording to play for me. The next day in the hospital, she played the tape of his little voice talking to me.

The doctors and nurses stayed in the room to watch this experiment, to see the monitors when I got to hear my son's voice. She played the message, and I began to cry. I was deep in the coma, but my body began to shake with emotion. That's when Carmen and our team of specialists knew for sure: from deep inside, I could hear them.

CHAPTER 13

On the day I woke up, my eyes felt like they had been scraped with sand. Everything was stiff, like my body had been covered tightly in plastic wrap. My eyelids were so heavy, and I wondered why I couldn't move. Shadows began taking shape, and I could vaguely make out the silhouettes of people in the room.

Someone said, "Hello, Sergeant Del Toro, do you know where you are?"

I opened my eyes against the bright lights to see a doctor standing beside me. I tried to speak, feeling for the first time the trach tube poking through a hole in my throat. "Afghanistan," I whispered.

"Sir, you are in the United States," the doctor said. "Do you know today's date?"

"December . . . something. Two thousand five," I whispered.

"No, sir. This is March 2006."

I looked around the hospital room, taking it in. I saw several people, all in masks and gowns and hospital garb. I recognized Chief Humphries, and I recognized my sister in the room. And then I saw my favorite sight in the world, my beautiful wife. I wanted to reach for her, but I couldn't lift my arms.

"You've been asleep for a very long time," the doctor said. "We're glad to see you, sir." He began to tell me what had happened since I was last awake. He doled out information slowly, one sentence at a time, sugarcoating nothing.

"Sergeant Del Toro, there was an explosion in Afghanistan. You've been badly injured, sir."

He took a breath or two between each sentence.

"Eighty percent of your body has third-degree burns. We thought you had only a fifteen percent chance of survival. You almost died on us three times."

I learned that I had a tracheostomy and that several of my fingers were gone. I learned that I was wrapped in bandages, and that's why I couldn't move anything on my body. I learned that I had been asleep in a coma for four months of my life.

"Sir, even though you're awake now, you still have a long recovery ahead of you. We predict you'll be in this hospital for a year and a half, and you may not walk again. Because of the damage to your lungs, you'll probably be on a respirator for a long time, maybe for the rest of your life. And, sir, while we thank you for your service, we have reason to believe that your military career is finished."

The doctor delivered that final blow of information to a silent room, as each person waited to see what I would say.

I couldn't speak louder than a whisper. The effort was great, and my voice sounded like my throat was filled with rocks. They couldn't hear me, and they wondered if I would give up in that very moment, overwhelmed by the great pain, the deep loss, and the hardest road ahead. But I assure you of my response. From the sandy grit in my soul, I thought, No, doc. You can go to fucking hell. That's not how this is going to go.

That awakening conversation was rough. The prognosis was grim, but it wasn't the first time someone tried to prepare me for a sad story they thought would be my life. It wasn't the first time someone tried to tell me how this would go. I've never accepted anyone's prognosis for my life, and I wouldn't start now.

I didn't accept it when I was a kid growing up on the east side of Joliet without my parents. When people said I should have been a gangbanger or a drug dealer, I chose neither. When I graduated from high school, when people said the best education I could get was junior college, I showed them how wrong they were with a full-ride academic scholarship to the University of Illinois. I had become accustomed to people projecting a sad outcome, and I plowed through the limits of their expectations. So, why would I now accept these doctors' diagnosis of me? I would not.

My last name is Del Toro. In Spanish, that means bull. The name fits.

A person's mind is a very powerful thing, far stronger than you can imagine. A small spark can light up the darkness, and so you must find your spark to light the fire inside you. If you believe you can overcome, then you will overcome.

And then you must hold on to that hope, keep that spark alive, every minute of every day. Nothing in the whole world is more powerful than a small spark of hope that refuses to give up.

In those first moments of my new life, I lay in that hospital bed, thinking of all my reasons to hold on to hope. I wanted the future I promised with that beautiful woman standing at the foot of my bed. I wanted to keep the promise I had made to my dad before he died. But most of all, I wanted to keep my promise to my son. He became my spark, and I knew I would do anything to keep that blaze burning for him.

There was one more flame of motivation blazing within me. I wanted to live, breathe, and walk on my own so I could show those sons of bitches who planted that bomb on the side of the road: *Fuck you. You did not ruin my life. You will not take this from me.*

"Sergeant Del Toro, we want to make sure you're not in any pain right now, and we can get you any medications you need. What would you like right now? What can we get for you?"

They leaned in close to hear my answer, my first request in those first moments out of the depths of my darkness.

I thought, Start strong, finish strong.

And I whispered, "I want a double quarter pounder with cheese from McDonald's."

The people in the room laughed.

The doctor said, "I hear you, sir. Maybe in a couple of days. You just woke up from a coma. It's important that you relax and take it easy. How about we start with a sip of water, sir?"

And so began my new life. I had to learn to drink before I could eat, to sit up before I could stand, to take a breath without assistance. Ultimately, I had to begin learning to live in a body that didn't feel like my own.

Before I left for Afghanistan for that last mission, my teammates and I were in a rigid physical routine of cardio, strength, and endurance training. Air Force standards require airmen to maintain levels of fitness, and I was at the top of my game. I had been a two-hundred-pound muscle-head, daily lifting weights as much as three hundred pounds.

But in those months in the hospital, my weight dropped to 115 pounds, and I no longer had the strength to even lift my arms.

You see, when a person falls into any kind of a coma, that extended time without using their muscles often leads to weakness, atrophy, and even permanent disability from the lack of use. In my case, burn injuries added to that list of risks. My body was covered in scar tissue that had grown thick like rope in some places, and stretched skin that was tissue thin in others. After the accident, my muscle mass had melted.

As the wound in my throat began to close, I transitioned to a smaller and smaller tracheostomy tube. Each one was a little more comfortable, though I never got used to breathing through and speaking around a hose in my windpipe.

As I began to breathe on my own, they gave me permission to start to stand and walk again. That was bittersweet news. Of course I wanted to walk, but the pain was unspeakable. The burns had gone into the innermost layer of my skin, damaging the muscles and tendons. When muscles, tendons, and skin begin to heal, they contract like tight rubber bands. To regain my range of motion, I underwent daily sessions of physical therapy with torturous devices to stretch me out again. The therapy was necessary, but the experience was similar to the tortures of war.

Physical therapy presents a different challenge for burn victims, because if you put too much strain on skin while it is healing from burns, the scarred skin can crack from the pressure. Even when I had the willpower to keep working, I had to give my skin a break. A team of nurses—including Carmen—changed the dressings many times a day, lubricating my skin and massaging my scars. Each time, the process both soothed and scathed. Their touch relieved some of the itching and aching, but to touch a burn is a pain that will make a grown man cry.

Burns don't kill people. Infections do. So, to keep from getting an infection, you literally get skinned alive. Without any skin to protect your insides, the room's temperature must always be kept at ninety-seven degrees to maintain your body temperature. Anyone who comes in the room must be covered from head to toe, and most people cannot tolerate that kind of protection in such a hot room for more than a few minutes. It was like an oven in my room. Add a full covering of mask, gown, and gloves, and it was broiling for any visitor. I felt comfort knowing someone was nearby, ready to help—or call for help—if I struggled to breathe. My visitors were relentless, in both presence and encouragement, but they couldn't stay with me for very long. Time in my room took a toll on them.

Nights were a living hell. There were no visitors allowed, and I was alone in a room with all my machines. I couldn't swallow, so there was a tube that would suck up the saliva in my mouth. But if the tube malfunctioned, or even if it subtly shifted to the wrong spot, my mouth, throat, and airway filled with fluid. I was awake, afraid I would drown. Many nights I lay in my bed, watching for the sunrise, hoping I'd see it one more time.

The nights were tolerable only with my two best nurses—Bonnie and Kim. They would come to check on me often, bringing me ice chips to cool my lips and quench my thirst. They made it almost okay, almost tolerable. But even the best nurses need a night off, and I hated all the nights without them. I was afraid I'd wake up dead.

But worse than the muscle work, worse than the nights I feared I would drown in my hospital bed—absolutely the worst pain of all were the therapy exercises to desensitize my skin. Burned skin becomes hypersensitive to touch, so the lightest brush of a feather felt like razor blades. Day after day, I withstood simple practices, like brushing a dry washcloth against my skin, dipping my hands in rice or sand, and rubbing my hands on carpet, rocks, or marbles. We had to progress to rougher and rougher textures, progressively teaching my skin how to respond. The pain was mind-numbing, but if I didn't do it, I'd never be able to hold even a spoon in my hands.

I'd never be able to hold my son.

I couldn't let that happen. I *would not* let that happen.

CHAPTER 14

People ask this question most of all: *Was there ever a time when you wished you had died in the explosion?*

I always say no, I never wanted to give up. But if I am brutally honest, there was one time. It was the darkest day of my life, the day I saw myself in the mirror.

When a person is severely burned, the hospital staff covers the mirrors in the room. The work of rehabilitation is grueling enough, even without seeing the ways your appearance has changed. They want to ease you into your transition to what you look like now. So, most of the mirrors at my hospital were covered, allowing us patients to gradually ease into the reality of our injuries. I imagined I didn't look the same, but I had no mirrors to tell me for sure.

I could look down at my body and see that some things were different. I knew I was missing some fingers, but I could see and feel that I still had my arms and my legs. I figured my face was pretty much the same. I considered maybe I had some singed hair, but I thought I'd look pretty similar to the man I once knew. You know what you look like in your mind's eye, and I thought I did, too.

One morning, I was walking to the restroom. It was a slow and careful process, with helpers on each side of me. My wife assisted me, along with Gary, my therapist—a six-foot-six bald white dude whom I called my guardian angel. They were walking alongside me, helping me to keep my balance, and we were almost there. But when we got into the bathroom, I lost my footing.

As I stumbled, they grabbed me to keep me from falling, and one of them accidentally pulled the towel off the mirror. Nobody meant for it to happen, but suddenly I saw my face for the first time since I had been blown up.

I gasped aloud to see another someone in the room with us, a monster with a mummy's body, a swollen head, and a disfigured face painted dark shades of red and black. Who was that monster?

Staring at the reflection, I made the connection. That monster was me. This is what I looked like now.

I broke down, weeping with angry rage.

"Why didn't you let me die? Why did you let me live if I look like this?" I begged them as I sobbed on the bathroom floor, wishing I were buried with my teammates who had gone before me. Gary and Carmen tried to calm me, but I was inconsolable.

I was a grown man, thirty years old, and I just scared *myself* in the mirror. If *I* think I'm a monster, what will my three-year-old think? The thought that my face might terrify him—that crushed me. It wasn't a question of vanity—that wasn't it at all. But I couldn't bear to let my son see what I had become. Guero was my strength, my spark, my entire inspiration. I would rather die than scare my son.

"DT, you can't give up, dude," Gary said.

"Please, just let me die," I cried. "I can't do this."

"No, DT," he said, "you can do this."

Gary got down on the floor beside me. He said, "Do you have any idea how many people you inspire every day? Every single day, man. You start the day fresh, asking people to get you up, get you stretching, and get you started on some of the most intense pain a human can go through. You inspire all of the staff, but it's more than that. You inspire the other wounded men in the hospital. They watch you overcome, and they believe they can, too. You're keeping people alive, DT."

"I want to die," I cried. "I cannot do this to my son."

Gary knelt down lower, placing himself in my line of vision. He spoke sternly, man to man. "DT, listen to me. Look at me."

I lifted my eyes to this bald white guy, my guardian angel.

Looking right into my eyes, Gary said, "DT, all your son wants is his dad back. That's all he wants. Trust me."

I couldn't imagine that my son could ever stand to see what I just saw. I broke down further at the very thought of it; it was more than I could bear. My greatest fear had rooted itself in the heart of my brain: The first time I see my boy, will he be terrified of me?

For nearly an hour, I was sitting in the bathroom, wishing I had died. It was the darkest day of my life. For one hour of one day, I fell into a dark pit of despair. Not from the pain of living in this body, but from the overwhelming worry of making life harder for my son.

The flame of my spirit almost went out that day.

But the spark of my son would not fade. If Gary was right, if his dad was all Guero wanted, then I would give him everything I had.

I stayed the course, moving out of the ICU and into a regular room. I worked through the agony of therapies so I could get stronger every day. In the afternoons, Carmen sat with me after my hours of therapy, exercises, and treatments. One day, she played that audio recording of Guero talking to me, just as she had played it while I was in the coma. But the sound of his little voice tore me up.

I raised my hand for her to stop. "Please, please no," I said. I asked her not to play it again. God, I missed that kid. I didn't want to hear his voice unless I could see his face, see his entire self, right there before me.

I spent my birthday in the hospital that year. I could walk on my own, but only for about one hundred feet before I had to sit down and catch my breath again. But one hundred feet isn't nothing, and that was enough independence for my team of experts to believe I could transition to a house on the Army base.

The Air Force worked with Carmen, as they arranged for her and Guero to move into a home at Fort Sam. The house was near the hospital and accessible to my many needs, specialists, and therapists. Finally, Carmen and Guero got to move out of their little hotel room that was barely big enough for the two of them. When I was released from the hospital, I could live at home with Carmen, and I'd come back for therapies every day.

On the morning of my release, Carmen and the nurses changed my bandages once more, and I got dressed to enter the world. Wrapped like a mummy, I put a baseball hat on top. I'd have to be so careful from now on in any bit of sunshine, as my skin couldn't withstand the risk of a sunburn. All anyone could see was my face.

Two months after I woke up from the coma, I left that hospital—one year and four months ahead of schedule. The day had come, my moment of truth, the greatest test of whether all of this was worth it.

We arrived at our home for the very first time. I remember my friends were there, my teammates, and a lot of my family. As we stepped through the front door, my wife called out to my son, "Guero, Papi is here!"

I heard the sound of his little feet on the wood floor as he came running through the house. He raced around the corner through the kitchen, running straight for me, but then he stopped in his tracks when he saw me for the first time. My heart fell like a rock in deep, dark waters. I thought, Oh, shit. I knew it.

The air stood still in the silence of the room. Everyone waited.

Guero tilted his head to the side, looking at me.

Brave and slow, he took a few steps closer.

He leaned in close, studying my eyes, searching for anything familiar to him.

And then, my boy whispered my name like a question. "Papi?"

"Yeah, Guero. It's me. It's Papi."

At the sound of my voice, he leapt into my arms. It was the most amazing hug I've ever had. It was the best moment of my life as a dad, second only to the day Guero was born.

Carmen was close behind, so afraid Guero would knock me down or hurt me. She had been monitoring every inch around me for months, and her protective instincts were on high alert. She said, "Guero! *Cuidado!* Don't hurt Papi!"

With Guero in my arms, I lifted my hand to tell her all was well. Nothing could hurt me now. I believe I said something very dignified like, "Back the hell off, woman."

Gary had been right all along. All my boy wanted was his dad back.

Guero didn't care what I looked like. He just wanted his dad to come home.

CHAPTER 15

The Purple Heart isn't an award that you put on your bucket list, hoping to receive it someday. It carries esteem and high honor—but also mixed feelings. The Purple Heart is the oldest medal in the military, created by George Washington and awarded to men and women who are injured or killed while serving in any branch of the armed forces. It means you got hurt and you're out of the fight—at least for a while, maybe forever. It means you're not serving with your teammates anymore.

While I was recovering from my injuries, commuting back and forth to the hospital for daily therapies, I learned about wounded servicemen in the burn unit who had been awarded a Purple Heart. Each award came with much fanfare and celebration of the highest respect, and word got around when active-duty personnel arrived to award a Purple Heart.

I began to wonder if maybe my injuries hadn't qualified for the award, since not every wounded warrior receives a Purple Heart. There is a list of criteria and a process for qualifying that depends on the wound, the medical treatments required, and the engagement with the enemy. Curious, I asked Carmen one day, "What's a guy gotta do to get one of those? I wonder if I'll get one."

"A Purple Heart?" she asked. "You have one."

"What? *I got a Purple Heart?* When did that happen?"

"President Bush gave it to you."

She said he visited BAMC just after New Year's 2006, days after my wife and her sister had raised their cups in their hotel room, hoping for better things to come in the new year.

"Wait a minute. President George W. Bush? He came here?"

She nodded, smiling gently at the memory she had of that day. "Yes, Junior. He came here. And he awarded you with the Purple Heart."

I lay back on the pillow and thought about the

words she had just said. President George W. Bush awarded me with the Purple Heart.

"Damn, I wish I could remember that," I said.

Carmen told me what had happened, how it all started when the hospital staff asked for her paperwork. (A lot of people were asking Carmen for paperwork on a daily basis. This alone wasn't out of the ordinary.) They asked to see her green card because they needed to do a security check. After they had run her background check and secured her identity, they told her why. A special visitor was coming: President George W. Bush.

The hospital staff explained that there would be strict protocol for her few moments with the president, and they laid out the etiquette and requirements.

- Who is invited: two family members per soldier.
- What to wear: business professional wardrobe. (Carmen did not have any business professional clothing with her in the States, so she had to buy slacks and a blouse. She borrowed her sister's purse.)
- What to do: it is permissible to shake hands with the president.

- What not to do: it is not permissible to hug the president.
- What to expect: be brief. The president is a very busy man with many people on his schedule today, so his staff will keep things moving. Do not expect a long conversation.

Carmen said she agreed to all these requirements. She had never met the president before, and she was prepared to follow the rules and meet the expectations of the honor.

But then the hospital staff mentioned one last thing to Carmen: "We will cover your husband from head to toe, to protect the president from the trauma of his injuries."

Carmen would not agree to this. My wife told them, "No. This happened. This happened to my husband, and it has happened to all of these soldiers who are fighting for their lives as a result of this war. It happened to the soldiers, and it happened to their families. Please do not hide that from our president. I want him to see the truth. I want him to see the realities of this burn unit. We are here with these wounded soldiers every day, and I want our president to see what it looks like."

The staff pushed back. Her request was against protocol. To have a soldier lying in his bed, his gruesome wounds exposed to the commander in chief—that is not how they do this.

But Carmen stood her ground.

She said, "No, this is not about protocol. This is about you wanting to hide how hard it is, trying to hide the damage that has been done. With all due respect to the president of the United States, I want him to see my husband."

On the day of the meeting, all the family members had to be at the hospital an hour before the president arrived so that the Secret Service could secure the area. There were Secret Service agents on top of the building, and there were policemen blocking the streets of all the surrounding city blocks. In fact, Carmen couldn't drive there or take the hospital shuttle—she had to walk to the hospital that day because all the roads were closed.

The families met in the waiting room of the ICU, and the president's staff called the names of each soldier's family members, one soldier at a time. Carmen said there was a long time between patients. When they called for my family, Carmen was escorted to my room in the ICU, and she waited for the president to enter the room.

Even the president had to be covered from head to toe to prevent the risk of any infection, so all anyone could see and recognize were his eyes. He came in the room wearing his protective gear. But, per Carmen's request, nobody hid my injuries.

The president is the commander in chief, the leader of the free world, and he has to make the most difficult decisions every single day. I do not believe that the decision to go to war in Afghanistan was the wrong decision, but it came at a cost to so many people and their families. On that day in my hospital room, and on many other days as he visited many other wounded veterans, he took it in. When President Bush came to my bedside, when he looked at my scorched body, he counted a piece of the cost. The president cried.

President Bush came to my wife, and she extended her hand to shake his, just like the protocol had stated. She said, "Hello, sir. It is nice to meet you."

But President Bush didn't want to merely shake her hand. He asked if he could hug her. When Carmen nodded her head yes, President Bush hugged my wife, and he cried. He said, "*Lo siento mucho, Señora Del Toro. Estamos aquí contigo.*"

For the first time since she had arrived at the hospital, Carmen did not need a translator. President Bush spoke directly to her in Spanish, with no intermediary to stand in between. In her own language, President Bush told her, "I am so sorry, Mrs. Del Toro. We are here for you. You are not alone."

That meant everything to her. To this very day, it means everything to me.

In a small ceremony with just a few people—my wife, my sister, Chief Humphries, and Chief Gonzalez—President George W. Bush awarded the Purple Heart to me. The protocols had said he would be brief, in and out, but President Bush stayed at my bedside for a half hour. Carmen said he stayed in the room longer than any of my family members were able to.

There was no team of photographers, no TV cameras, no publicity, no featured reports on the news that night. The news channels reported that President Bush had been in town, but they did not report where he had been, whom he had visited, or why. He wasn't there for publicity and optics and interviews. He was there for the military warriors and their families, that we may know we are not alone.

You may not agree with his politics, but you can

never say he didn't care for his soldiers and their families. I can tell you that firsthand. (Well, Carmen can. And I will forever wish I could remember it.)

Carmen kept the Purple Heart with her for all the months I was asleep in the coma, hoping to show it to me one day, praying I would wake up to see the gift I had received. The president's staff sent her a letter, the private photos of the visit, and a coin to commemorate the day and her role as a military wife.

In June 2006, a few weeks after I had come home from the hospital, Chief Humphries told me, "DT, we have decided to repeat the ceremony for your Purple Heart. You deserve a ceremony you can remember, so we're going to do it again." Chief Humphries and my teammates planned a ceremony in the theater of Brooke Army Medical Center.

By then, I could walk very short distances. I rode in a wheelchair to enter the room, but then I walked onto the stage that day, short of breath and with assistance. I looked around that auditorium to see that there was standing room only. I met every surgeon who had worked on my body, and I saw every nurse who had helped me through the days and nights. I saw the faces of my brother, my sisters, the hospital staff, and even other guys who were in the hospital, wounded in the same war, fighting a battle of their own. My teammates had flown in from all over the world, and each of them stood together in a line, shoulder to shoulder, as I took the stage. And of course, there was my best buddy, the scrub, Jeff Mariano. He didn't let a month separate us.

Chief Humphries welcomed me, and he said, "DT, we have a surprise for you."

With a sweeping arm and an open hand, he gestured to the door of the auditorium. That's when I heard the sweet, familiar jingle-jangle of a dog collar. My teammate walked in the room with Dakota, our Golden Labrador, prancing on a leash.

Oh my God, I hadn't seen that sweet puppy for months, since Carmen and I left Italy. My guys in Afghanistan had worked together to bring Dakota to me, all the way from the army base in Italy. When she started wagging her tail, the whole back end of her shook with joy. There wasn't a dry eye in the whole room.

In case you've never brought a dog across the world, it's not a cheap adventure. A dog is expensive cargo, and it cost quite a bit to get Dakota here. My teammates had partnered with Carmen to bring Dakota to me, so she knew all along about this surprise. There were some great questions about whether we could have a pet, due to the risk of infections, but the doctor said she would be a huge help in my recovery. A family dog is a therapy all its own.

They kept her on a leash until after the ceremony, and would you believe this, she sat like royalty. She was quiet, she didn't bark, and she listened as if she was taking it all in. Dakota sat at full attention as I received the award from General Michael Moseley, the chief of staff of the United States Air Force.

General Moseley invited my family to join me on the stage. Carmen, Guero, and my brother and sisters came up to the stage, and General Moseley honored each of their sacrifices as military family members. It was a cool moment . . . until my son slapped my sister right on her ass. To this day, I have no idea what he was thinking, but he broke the ice. Everyone who'd been crying now broke into laughter, even General Moseley.

After the ceremony, we went back to the house for a barbecue that my teammates had arranged. That's when I finally got my hands on Dakota, to pet her downy head and smell her dog breath once more. That beautiful girl, she had missed us as much as we missed her.

As if that wasn't enough, my teammates had one last surprise for me: they had found my lucky hat. I had worn it every day of every deployment, and on the day of the incident, I had tucked my lucky hat into the left pocket of my cargo pants. It was the only thing on my body that had truly survived. Somehow, the hat had only a singed spot.

My guys found my hat in the wreckage that day. They brought it home from Afghanistan, tucked it safely away, and believed with all their hearts I would live to wear it again.

Obviously, they made me cry. Those assholes.

Dakota lived a good many more years with us, that brave girl with a heart of gold. The Purple Heart and my lucky hat are both in my home office. They represent the old DT, before the injury. I'm a different person now. I might have gotten a bad break, but I've still got a lot to live for—with a good family, a good dog, and a whole lot of good luck.

CHAPTER 16

My recovery was just beginning. In the years to come, I had more than one hundred surgeries all over my body, especially to encourage the mobility of my hands and to reconstruct my nose. A lot of me was broken, and my whole body was under reconstruction. I had let go of the idea of what I would look like, and I didn't have the surgeries for my own psyche or ego. But I wanted to have them for my son's spirit. Life would be hard enough for him, and I didn't want my son to have to protect the honor of a disfigured dad.

Carmen became an expert in wound care. Before I was injured, she was afraid of needles. She hated hospitals—even an appointment with the dentist made her squeamish. But when I was in the hospital, there was no time for that kind of weakness anymore. After several weeks of being in the coma, the nurses began to prepare her for what would happen if I woke up. They began to prepare both of us for our life outside the burn unit.

Carmen remembers when they said, "We need to train you, Mrs. Del Toro, so you will know what to do."

"But won't you be here?" She looked at me, then deeply asleep and intubated, and she couldn't imagine them shuffling us out the door when I opened my eyes. There was surely a long road ahead, and it had to involve a team . . . right?

They said, "Once he wakes up, we will move him to another unit for the next stage of his recovery. He'll stay in the hospital until he is well enough to leave, and eventually, you will take him home with you. He will only have to return to the hospital for a couple of hours each day for physical therapy."

"But who will change his dressing? Who will care for his wounds?"

"That is you, Mrs. Del Toro. Let us teach you what to do."

It would be Carmen. All Carmen. Only Carmen.

They invited her to watch as they changed my dressings, showing her how to use each tool. Before we left the hospital, they gave Carmen her own kit of surgical tools, cleansers, antiseptics, and dressings for wound care. She would need to change all my dressings twice a day, in the morning and at night.

I have had 150 surgeries so far, and Carmen has been there for all of them, through the healing and recovery for each one. She became my nurse. She does not have a degree, and she does not have a certificate to show what she knows how to do, but the nurses in the ICU taught her an honorary master class. She has become an expert in her field.

Over the course of so many surgeries, Carmen and I both learned very much about the skin graft, where the doctors take healthy skin from one part of the body and transplant it to replace skin that is damaged or missing. The hope is, within a few days, the grafted skin begins to develop blood vessels and connect to the skin around it. But this does not always happen. The transplanted skin does not always heal to the surrounding skin. When it doesn't adhere, we have to begin again.

To help the transplanted skin adhere to the healthy skin, the doctor makes tiny holes or several crisscrossed cuts in the transplanted skin, to make the skin look like a fishnet. Then they stretch the skin over the larger area of damaged skin, securing it in place with stitches or staples. They cover the new skin with dressing to keep the wound protected, and then the healing process begins. Sometimes it works, and sometimes it doesn't.

One time, as I was healing at home after yet another skin graft procedure, I felt an itch on my lower back. I couldn't reach it, so I called Carmen to scratch it for me.

"Sure, honey," she said. "Let me look at it."

She lifted my T-shirt and gently peeled back the dressing to see what might be causing the itching. She found a giant red sore, the size of a nickel, with a white circle in the middle of it. It looked like an acne pimple—but much bigger.

"Oh, you do have something going on back here, Junior," she said. We had to be very aware of infections of any kind through every stage of my recovery. Any infection, if left untreated, could become serious very quickly.

She got her kit of surgical tools, and she disinfected my skin. I felt the cool swab of alcohol on my back.

Carmen grabbed the special tweezers they gave her, a long, pointy pair used for surgeries, and she popped the little tip of this pimple on my back. She squeezed the pimple to drain the white pus, and then she cleaned it away with alcohol and bacitracin. As she squeezed the spot once more, she heard a clicking sound against the tweezers. It sounded like metal on metal.

I felt it at the same time. "Ow! Ow!" I said, flinching away from her. "What are you doing?"

"There's something there, honey. I think it might be a staple. We should have someone look at this."

"Take it out," I said.

"*Take it out*?" she said. "No, no. This is a job for a nurse at the hospital. Let's go to the hospital—"

"Carmen," I said, interrupting her. "Take it out. You can do this. Just be careful."

"Are you sure?"

"I am sure," I said.

I focused on my breathing, preparing for my wife to reach inside my back and pull out a piece of metal. She followed my lead. She took a breath of her own, and she prepared to do a minor surgery on me.

Carmen slid the tweezers inside the pimple, feeling for the metal staple. She poked softly against the piece, until a dark tip of stainless steel emerged from deep within my skin. She grasped the end of the staple with the tweezers, wiggling and tugging as gently as she could, until it slid out of my skin.

It hurt like hell. I groaned as she pulled it out.

Carmen nearly passed out. The staple was in her hand.

When we went to the doctor the next day, we explained what had happened, what she saw back there, how she removed it, and how she cleaned the affected area.

The doctor looked at the spot on my back, and he said, "Looks good, Carmen. You did great. Looks like a medical professional has been here. You did good."

Needles don't bother her anymore.

There is one thing I'll never overcome, one scar the explosion left behind that all the courage and willpower in the world will never recover. It's a heartbreak that many wounded veterans face, and nobody's talking about this wound so many of us have.

I come from a big family, a giant family tree with dozens of cousins, aunts, and uncles. I wanted to give my son brothers and sisters; I wanted to give my wife a daughter. We wanted to have more children, to grow our family. But we couldn't get pregnant.

Finally, we saw a fertility specialist who could diagnose the problem. His results showed that my wife's fertility seemed likely, and her body showed every indication to function as it should to carry another child. But the doctor asked me if I had had a vasectomy.

"Of course not," I said. That's when I learned that the biopsy and test results indicated zero sperm. The explosion had been my sterilization. There would be no more children in our family. It's a sadness I've rarely spoken aloud, one that has broken my heart many times over.

I say it here on behalf of the men who have made this silent sacrifice, in the radiation of battle. When we talk about what's taken from us, this is one we can't get back. And it's too painful to talk about.

The demands on our families are truly unspeakable. One day, I was finishing physical therapy at BAMC, and one of the volunteers at the therapy center was walking me back to the waiting room. I heard people shouting in one of the rooms nearby. It wasn't the sound of someone working hard through therapy; it was the sound of angry people, arguing over one another, bickering, and fighting to be heard.

"Damn," I said to the nurse. "Those people are getting after it. That's hard to listen to."

The volunteer said, "Yes, sir. Those are some family members fighting over the care of a service member. It's a very hard time for the families. We hear that quite a bit around here."

I thought of my brother and my sisters, how close we were, so committed to each other after all we had been through together. I said, "I'm glad my family didn't go through that."

She looked up at me, and she said, "Sir, I'm sorry to tell you, but your family sounded exactly like that."

"No way. Like those people? Raising hell in there? That's not my family."

She nodded. "Yes, sir. That could be your family. They fought just like that."

That was the first indication of the conflict that had torn through them while I was in the coma. I had had no idea.

Carmen never wanted to tell me about it. When I asked her, she broke down crying. She had decided that would be her secret, her burden to bear. As they argued and disagreed with her, she reminded herself that the only thing that mattered was my recovery. She only wanted me to get well and live. That's all she needed. She convinced herself it wasn't important to tell me about all the drama with my siblings, because I wasn't awake for it anyway.

Well, I was awake now. I was pissed.

I called a family meeting, bringing together my siblings and my wife. I told my brother and my sisters, "Listen, Gueris didn't want to tell me about this, and I had to hear it from a volunteer at the damn hospital. But I know now, how you guys didn't support her, how you fought her while she was making decisions on my behalf. I thought my family was better than that. I had to learn from a stranger that you didn't have my back, you didn't have my wife's back, and that—as a family?—we were a mess."

I said, "From this day forward, if you want to be in my life, then you have to respect my wife and my son. They are my family. You guys are my brother and sisters, but you don't come first anymore. You're second. I'm sorry to say it, and I love you. But I better not hear, ever, that this happens again. You better have my back from this day forward. And that means you have hers. Don't make anything harder than it already is. You want to love me well? Support my wife."

Yes, I had become like a father to them, and I was ready to take the lead once more. I would lead my family through this, first of all, by taking care of my wife.

CHAPTER 17

In the Air Force, we pride ourselves on being the best at taking care of our people, and that's typically very true. The benefits of military service are a big deal.

Military members and their families have a health insurance plan that covers medical, dental, and vision, so our medical coverage is basically free to us and our families. There is inexpensive life insurance available to trainees when they begin boot camp, so they know their families will be taken care of should anything happen.

While on active duty, housing is always inexpensive and sometimes free, and all relocations are covered by the military as well, as long as the military is requiring you to move. Family members have access to all of the on-base services, including amenities like the gym, swimming pool, and commissary. The commissary is the grocery store located on most bases, and it looks and functions like a regular grocery store—except the prices are about 30 percent lower. The BX is the base exchange, which is like a Walmart or Target, complete with everything your family might need. They match the prices of any competitors in the area, all without sales tax.

There are great education benefits, like tuition assistance while you're enlisted, so you can take as many as six classes per year while you're in the Air Force, completely free of charge. Basic training and tech school count as career development courses that can be transferred to most universities, and of course there is the GI Bill which provides money for any individual who has served in the military to attend college.

The leave and sick days are generous as well. An active-duty Air Force member will accrue two and a half days of paid leave per month, for a total of thirty days paid vacation per year. Paid sick days are unlimited, provided nobody's calling in sick with a hangover after Thursday nights at the Broad Street Bar. That is grounds for a dishonorable discharge, so don't mess around with that. But if you're sick, you're sick—with no penalty.

And the retirement benefits are famously good. If you serve twenty years in the Air Force and get out, you immediately begin receiving 40 percent of your last paycheck for the rest of your life, and that's even adjusted for inflation each year.

So yes, without question, the Air Force takes good care of their people.

But when it came to our wounded, injured, and ill, the Air Force was completely outdated and severely behind. There was great care for pilots, but I was the first of my kind for the Air Force—the first to be injured by a roadside bomb. Not many airmen are injured the way that I was, since most of them are in the sky. When I got hurt, they didn't know what to do. They're not used to someone like me surviving.

Here's something you may not know. As burn patients, we often feel like second-class citizens, especially compared to the amputees wounded in battle. Listen, we know what we look like. We know we aren't as easy to look at as the guys whose faces are still pretty. We know we're not as interesting as they are, with their robotic arms and state-of-the-art prosthetics. But the conditions should be equitable for every wounded warrior. And when I was recovering, that was not the case.

The prosthetic patients had a giant facility for their

rehabilitation. They had the most intelligent biomedical equipment, specially designed for their recovery. In the burn unit, we had twenty burn victims doing therapy in one locker room, and our only piece of equipment was a bike.

The wounded veterans of the Army, Navy, and Marines had case managers, social workers, and a whole wing for recovery. Their community was staffed with caseworkers to navigate the medical process alongside them. At Brooks, I had no case manager. Lucky for me, I had Chief Humphries and Chief Gonzalez who stepped up to help me, but most soldiers didn't have people to advocate on their behalf. In the Air Force, we didn't have that team, and in the burn unit, we didn't have that support.

It's a discouraging discovery, after you've fought so hard to save your own life, to realize they never really expected you to live. There wasn't a plan in place for those of us who survived.

So, I did the next thing to be done: I became a patient advocate. I started going to the hospital, visiting one guy at a time. I talked to each one.

I said, "Man, I know this is hard. I know how hard it is. I was there, too. But you've got to find your spark, man. Find your drive. It can be anything. You may not think you have anything to live for, but I promise, you do. I won't tell you that there won't be hard days—I had hard days, bad days, too. But don't let them keep you down. I'm here for you. We are here for you. We are your team."

I wanted to give back to them, these people who needed it. I wanted to believe in them for real—not just tell them they could do it, but also meet them on the other side when they actually made it there. We've got to meet them there with a plan for what to do next. When there's a plan in place, it means your people really thought you were worth investing in, believing in, and fighting for.

It means they really thought you'd *live*.

In 2007, when Brooks Army Medical Center built the Center for the Intrepid, they designed a four-story medical facility designed to be used for rehabilitation, research, education, and training. They professed to accommodate the amputees with the most sophisticated care, with challenging sports equipment and virtual reality systems to fit and fabricate their prosthetics.

Which was great. But they offered nothing to burn victims.

Meanwhile, they had the audacity to invite all of

the burned warriors to their opening day and ribbon-cutting ceremony. Why? Because burn patients outnumber amputees three to one. They wanted the optics of our support and presence, but they weren't offering us a fraction of their services.

So, I put the word out to my brothers and sisters. I said, "Guys, we are not going to go to the ribbon-cutting ceremony if they won't let us be patients inside that center. We aren't going to fill their ceremonies with pomp and circumstance. We don't have to be the faces of their mission if they do not intend to help us."

Word got out that my colleagues and I weren't so eager to attend. Suddenly, there were conversations about a big meeting with staff at the hospital, and—wait for it—they wanted to speak with us. For the first time, the burn victims had a voice at the table. I am glad to tell you, from the day of that meeting on, the Center for the Intrepid included services for the treatment, education, and rehabilitation of burn patients.

As I began to use my voice as a wounded warrior, there were people in the Air Force who supported me, helped me, and started asking what we could do better. With the help of many others, I was able to implement change in a lot of outdated policies for wounded, injured, and ill service members in the Air Force.

Formerly, a service member who was put on medical hold could not make rank, test for rank, or be submitted for medals for achievement. That's one policy I helped to change, even though I didn't get to benefit from it.

I worked closely with Mr. John Beckett, the Washington-based program manager for the Air Force's Wounded Warrior and Survivor Care programs. Mr. Beckett was an integral part of changing policy, and the program brings comprehensive services and programs to the newly injured and their families who are rebuilding their lives. The Air Force Wounded Warrior Program (AFW2) was built on my suggestions and input on behalf of men and women like me.

The Air Force now has a family liaison officer program in which a team is appointed by local commanders to provide assistance to injured military members and their families, as well as the family members of those who have died. The family liaison officers (FLOs) support the families with medical appointments, housing, transportation, financial issues, daycare, and any additional needs they may have. They take the stress off the family members, so those people can focus on their loved ones.

Essentially, they are the angels Carmen needed while I was in a coma. That team is in place now, bridging the gap between what has happened and what must be done. Every warrior deserves a chance at life again, and I am honored to give a voice to the airmen with visible and invisible wounds.

In my recovery, I missed being with my teammates. I missed being on the range, but I knew that these guys in the hospital were my teammates now. Yes, I was wounded, burned, and in recovery, but I had been trained as a noncommissioned officer in the Air Force. An NCO takes care of his guys, and I had a job to do.

Your job as a leader is to make things better for those who come after you, even though you may never see any of the benefits. So that's what I started doing: changing policy to better the lives of our teammates.

We have to show them they're still our teammates, we're still fighting for them, and we still believe they've got this. We believe they will live, and we have a plan for what's next.

CHAPTER 18

In January 2009, Carmen and I received a letter from the White House. President George W. Bush was leaving office at the end of his second term, and he invited us to his farewell address on that last night before President Obama would take office. President Bush invited us to be his guests of honor at the event. I dressed in my full uniform for the chance to finally meet my commander in chief and the first lady, the lovely Mrs. Laura Bush.

When we came through the line to greet him, he knew my wife already, and he knew my son. He spoke to Carmen in Spanish, checking in to see how she had been since they last spoke. I shook his hand and he joked, "DT, good to see you again. You look a lot better than the last time I saw you."

Conscious and vertical and breathing on my own, I could definitely agree. "Yes, sir. A little better." This time, I was awake in his presence, and I would never forget it.

We stayed in touch over the years, and I have had the honor to meet other presidents as well—including President Obama, President Trump, and President Biden, back when he was vice president of the United States.

But I have to say, President Bush has even become a friend of mine. One day while I was at the gym, my phone rang with a phone number from Texas.

"Hello, is this Sergeant Del Toro? This is Freddy, the chief of staff. President Bush wants to talk to you. Hold one moment, please."

Well, this was a first. A moment later, I heard that familiar voice: "DT, it's George W. Bush. How are you today?"

"I'm doing well, sir. Just working out at the gym."

"Yes, I would hope so. It's about time you get to the gym," he said. He likes to talk smack, and he knows I can handle it. That became the first of many phone calls, and he often calls my cell phone in the middle of a regular afternoon, just to bust my chops. I dish it right back.

I said, "Sir, anytime you'd like to meet me at the gym, we'll go at it. You and me."

In 2017, after leaving office, President Bush released his book *Portraits of Courage: A Commander in Chief's Tribute to America's Warriors.* He wrote the stories of the courage and resilience of military veterans, and he included sixty-six full-color portraits of veterans, each one painted by his hand. With his own passion and inspiration, he featured the stories of the United States military, each of whom he has come to know personally. What an honor to be on his list, to be included among his soldiers.

When he did the painting of me for his book, President Bush invited me to go with him to New York for the *Today* show, and then to fly back with him to Tampa to meet all the SOCOM guys, the Special Operations Command. As we were flying back to Tampa on his jet, he started giving me a hard time.

"So, DT, tell me. Why did you leave Texas? Why'd you leave God's country?"

"Sir, I'm still active. I report to you. If you didn't

want me to leave Texas, then you should have done something about that."

He just kept busting my chops, until I finally said, "You know what, sir? I'm going to kick your ass."

George W. Bush busted up laughing. We all did. Freddy turned to the Secret Service agents, forever present in the room. He said, "Did you guys just hear that? DT said he's going to kick his ass. Seems like a threat to me. Seems like you guys should do something about that."

But the Secret Service agents waved it off. "It's all right. They can handle each other."

Have you ever had the Secret Service say, "Eh, it's fine. He's not a threat"?

Because who tosses a threat like that to a former president? You only say that if the Secret Service agents know you so well that they know your intent, if you know they aren't going to pin you to the wall. You only test those waters if you can be sure you won't drown.

I'm still swimming—and running, lifting, and working out. And it will be my privilege to run drills with our former president anytime he's ready.

CHAPTER 19

Many of us wounded soldiers are also former athletes. We played sports in earlier seasons of our lives, and it was part of who we were. When you lose a limb, or when you're severely wounded, you can't imagine you'll ever play sports again. How could that even be possible? We have lost our ability to play, and that often means we've lost access to a whole aspect of society.

Most people don't realize it, but sports and athletics are an entry point into community and society. It starts when we're little kids, playing Little League and flag football. Children learn how to play a game, but they also get to be part of a team. Sports build self-esteem, confidence, healthy competition, and teamwork. This grows and deepens as we get older, expanding into teenage and adult recreational leagues for softball and indoor soccer. You don't have to be a professional athlete to enjoy the social merits of the game. But when you're wounded, that option feels like one more thing taken away. I thought my sports career was over.

It's tempting for burn patients to isolate. We know we don't look the same anymore, and we're afraid people will stare—because they do. Kids will ask questions, like, "Why is your face like that? Where are your fingers? What happened to you?" Adults have the same questions, but they don't have the bravery to voice them out loud. Instead, they will silently stare. So, it takes courage for us to enter the world again, and a whole lot of us are hesitant to get back out in society. It's just too hard.

As I worked with the Center for the Intrepid, my therapists began to introduce me to adaptive sports, a world I had known nothing about. It's a whole subculture, and especially in San Antonio, there is a very large adaptive sports community. This helps with physical recovery, yes, but adaptive sports are also a path back into society. With adaptive equipment, virtual trainings, nutrition plans, and expert coaching, we discover that the world of athletics isn't over. Not only are there sports for us, but there are people like us in the game.

My first adaptive sport was sitting volleyball, and it's exactly how it sounds. You're sitting on your butt, playing volleyball. No wheelchair, just straight sitting on the floor, playing in a smaller court with a lower net. It's very fast-paced, and you stay in a sitting position at all times, using your hands and arms to maneuver around the court. Sitting volleyball helped me to restore my range of motion, and it helped develop my dexterity with my hands. It became part of my recovery, and I started to excel. With an entry into society, I began to feel like myself again—and I had really missed that guy.

I also tried adaptive cycling, where the para athletes compete on handcycles, recumbent bicycles, or tandems. Then I tried adaptive track and field. I trained in the throwing events, including shot put, discus, and javelin. With each one, I discovered a part of myself that still lived inside me, stronger than ever.

As I began to excel, I attended a camp in the Paralympics, a Military Adaptive Sports program in Chicago. They were hosting a powerlifting event, and I went to support my teammates who were competing in that event. Powerlifting had been my passion before I was injured, but I was sure I'd never be able to lift free weights again. I had been working out on machines for years by then, but I thought free weights were a hobby of the past. My hands couldn't do it anymore.

Mary, the Paralympic powerlifting coach, came to me on the sidelines. She said, "DT, I think you should give powerlifting a try."

"There's no way, Mary." I laughed it off to hide the heartache. "I can't do it anymore."

"Sir, I've seen what you've done. I've seen you do some great, amazing things, and I know you can do this."

I shook my head. "I don't know, Mary."

She pressed on. She said, "A lot of these athletes and veterans look up to you. They know the odds you were given, and how you've overcome. If they see you can do this, they'll believe in themselves, too."

Those words will forever inspire me. If my efforts can inspire someone else to try and maybe succeed, then I'll give it everything I have, every time.

"We'll spot you until you're ready," Mary said. "You should give it a try."

With a challenge like that, how could I not give it a try?

Para powerlifters lie on their backs on a bench

press, strapped in for this adaptive sport. So, I lay down with my back flat on the bench, and they strapped me in. I placed my hands under the bar, and that's when I discovered that they fit perfectly on the bar. There are many things my hands could no longer do, but their new shape seemed to be specially made for this unique task.

They put 45 pounds on each side of the 45-pound bar, 135 pounds in total. They have a spotter on the side of each powerlifter, just in case the lifter needs help, and they'll hold it until you say you're ready.

"Got it?" the spotters asked.

"Got it," I said.

"When you're ready, say go."

I exhaled the air from my lungs, thick with scar tissue and fierce with determination. I said, "Go."

They said, "Lift."

I brought the barbell slowly down, breathing into my chest. I held it for two seconds.

They said, "Press."

I fixed my eyes on a spot on the ceiling, then I pushed the bar up to where it belonged. I locked my elbows and held the bar steady and controlled, waiting for the signal that my lift was good.

It was *good.* I had done it.

The spotters put the bar on the rack, and I broke down in tears. I never thought I'd be able to do free weights again, not ever in my life. I thought it was over for me, but I had done it again. I can't describe it, that rush of emotion and victory.

Knowing what that felt like, I was ready to try anything.

When I got home from that trip to Chicago, my team introduced me to air rifle shooting. Para athletes use the same ranges and targets, and the competition format is the same. The target is comprised of 10 concentric scoring rings, and you want a perfect shot, which is 10.9. That's dead center, the perfect shot that you're aiming for. The bullseye is only about 0.05 centimeters wide—about the size of the period at the end of this sentence.

When you're in track and field, throwing, or even in the weight room, you need an adrenaline rush to launch the shot put, the javelin, or the disc. You listen to loud music to hype yourself up.

But shooting is a whole other monster. With shooting, you have to keep your heart rate down. You have to relax and stay calm. You get sixty shots in an hour, so one minute per shot. That may not sound like very much, but it's tiring. Your eyes get fatigued. Your body gets tired, because you have to keep that same position every time. Your breathing has to stay even. It's a whole different discipline.

My team adapted an air rifle so I could shoot it. First, there was a little tube for me to blow into to shoot, and then they were able to adapt the trigger so I could shoot with my finger. Since I can't load my pellets, I have a loader: my son. Guero would load my rifle. By this time, Guero was about twelve years old, and Carmen and Guero and I were an unstoppable team.

When you're shooting, you put in earplugs because you want to zone out. You don't want to hear anything that can distract you. I couldn't hear very much while I was shooting, and that was by design. But every time I got off the shooting range, my teammates would tell me how my son had cheered me on. They were laughing and giggling about how funny he was. They said he's like a little coach, whispering to me, "Come on, Dad. You've got this. Stay strong, finish strong."

They are the words that I said to myself on the first

day I woke up from the coma, and they have become our motto. Now that he is a grown man, Guero and I have matching tattoos with those words: Stay strong, finish strong.

CHAPTER 20

I never forgot the image that flashed before my eyes on the day of the explosion: my wife dressed in white, a bride in a church. I still wanted to give my wife that proper wedding. We had been officially married all those years ago, but I wanted to give her the celebration.

I wanted everything we had intended to do from the start, when our plans kept getting canceled or delayed by my deployment. I wanted everything for her—the white dress, the bridesmaids and groomsmen, the traditions in the church. We kept planning a date, but then I would have to go downrange, and we would have to reschedule—again. But the promise of that wedding had kept me alive, and I wasn't going to let it pass me by. Finally, in 2009, we could make it happen.

With our family spread across Mexico and throughout the United States, our first task was to narrow down where to get married. We decided the most central location for all of our people would be in Chicago, and we chose to get married at the Cathedral of Joliet, with its classic wooden pews and regal high ceilings. Carmen had her dress made by a family seamstress down in Mexico, and we chose all the people who would stand in our wedding with us. A giant wedding party with my teammates and family members from every part of our family tree.

We had one decision left to make: where to have the wedding reception. It was an essential decision, because you have to know this about us: Del Toros know how to throw a party.

I reached out to Anna Davlantes, the cohost of *Good Day Chicago* on Fox 32 Chicago. She had done a couple of stories about me as a veteran from the Chicago area, and she had become a good friend of mine. I knew Anna would know the best places in Joliet, the best venue to host one hell of a party. She told me about a country club in one of the south suburbs of Chicago. She was a member there; she could put in a good word for us, and she confirmed they had the date open . . . and they offered a military discount.

Say no more. We booked it. My bride and I were getting married.

Next, on a wing and a prayer, I reached out to Richard Marx, just to see if he might be able to perform "Right Here Waiting" at our wedding reception. He had a concert booked for that night—but he sent me the next best thing: he recorded a special video of his band singing to us. It would be our first dance at the reception, a surprise for Carmen.

I was on active duty, so I stayed in Texas right up until the wedding, but Carmen and Guero drove up to Chicago almost two weeks before the day to make sure things were good to go. As Carmen was tying up all the details, confirming the flowers and the menu and the mariachi bands, I checked in with the country club to see about paying the bill. I just wanted to know how much this was going to cost me.

They kept putting off that detail with me, giving me some reason to wait. They needed final numbers for the guest list, or some other detail. I was starting to get a little nervous. I mean, we were down to the wire, and everything was booked and underway. We had two hundred guests coming, and I knew this was going to be a bill to pay. I just wanted to be prepared, but I would have to wait and pay at the end of the reception, when it was all said and done. I didn't want them to hand me a bill that would blow me up all over again. (Guys who have been blown up get to say things like this.)

The ceremony was beautiful in every way. We invited Father Jimmy, the hospital chaplain from our days at BAMC, and he officiated the wedding along with my cousin Philip. We had a mariachi band there in the church, just like at a wedding in Mexico. It was perfect.

After the ceremony, we went to the country club for the wedding reception. All the groomsmen and bridesmaids came into the ballroom, then Carmen and me, and we were seated for dinner. It was almost time for the first dance. Anna Davlantes stepped up to the microphone, and I knew she was going to introduce the Richard Marx video. She said, "I have a surprise for you, Carmen, but I also have a surprise for DT."

I looked to Carmen . . . what was she up to? But my bride didn't know anything about it, either.

Anna said, "The first surprise—the surprise that neither of you know about—is this: the members of the country club are paying for the whole reception."

You guys. That was twenty grand.

The country club members came out to the dance floor, and I went down the line, shaking their hands. I thanked each of them for such a remarkable gift.

And then they lowered the screen in the ballroom, and Anna said, "And now, for a surprise that Carmen didn't know about. Carmen, we have a special message for you, from Richard Marx."

Richard started talking on the screen, and of course he started busting my chops, asking Carmen how she could handle marrying me, questioning her judgment, making the whole room laugh. He said, "I wish I could be there to celebrate, but hopefully this is the next best thing."

The camera panned out to show his whole band, and they started playing those familiar chords, the opening notes to "Right Here Waiting." I invited Carmen onto the dance floor for our first dance. Carmen immediately started to cry, and then everyone in the room started crying.

My brother-in-law only knows a few words of English, and he certainly couldn't understand the words of the song, and even he was crying. Later he told us, "I didn't know what they were singing, but I knew it must be good. I saw that everybody else was crying, so

I started crying too." That's how you know it's a good surprise: when people don't even know why they're crying. That song will do it, too.

We finished our first dance, and then the guests joined us on the dance floor. That's when the party really started. My teammates were my groomsmen, and they bought personalized cups for the entire wedding party. These weren't just like a pint or a beer mug . . . these were like the size of a jug. This giant cup was bigger than the palm of my hand, hollow all the way down the handle, each with our names on them.

My cup said "Pimp."

They kept that cup full all night long. Anytime I got close to finishing it, somebody filled it up again. I had a blister the next day from carrying it around. Some people get a hangover from drinking alcohol, but I got a blister from my giant jug of beer.

We got to the end of the night, and one of my buddies had a great idea to finish off the party. Jeff said, "Hey, DT. How about a little throwback? I got a CD in the car. Let's bring the Backstreet Boys back one last time."

Just like our dollar-beer and karaoke nights at Fort Bragg, my boys and I took the stage. "*You are my fire, my one desire.*" It was all just like old times, only this time, I'd gone through the fire to sing about it on the other side.

We started the night with Richard Marx, and we wrapped it up with the Backstreet Boys. It was silly and ridiculous, and it was magical—all of my closest friends, all these muscle-heads, dancing in a circle with my wife in the middle. We celebrated a day that almost didn't happen. That's what I wanted to give to Carmen.

By the way, Jeff the scrub finally got married, too. He put his wedding on hold, even though I never wanted him to do that. I asked him to live his life, to get back in the game, but he said he would wait to get married until I could stand at his side. Even when I could walk again, I was in the middle of a series of reconstructive surgeries to fix my nose, and my face was a mess with scars, tissue, tubes, and drains. I told him, "Dude, you don't want this ugly mug in your wedding photos. And I'm pretty sure no bridesmaid wants me to walk her down the aisle."

Do you want to know what that scrub said to me? He said, "I stood in your wedding, and you're standing in mine. And if a bridesmaid doesn't see the honor of walking with you, then she can walk herself right out of my wedding." I'm telling you, the brotherhood is fierce.

As for the third vision from the day of the accident, Carmen and I still haven't honeymooned in Greece, though we've seen a lot of beautiful places. Maybe someday, but I'm content with these dreams come true.

After all, it would be hard to top that night with the Backstreet Boys.

CHAPTER 21

General Moseley, chief of staff of the Air Force, invited me to be on a speaking panel to talk with military personnel about my story and what had happened to me. Moseley said, "DT, I have this panel that goes around talking to other service members. I would like you to be part of this speaking tour."

I balked at the very idea. "Sir, I'm not a speaker. I'm an operator. I'd rather be downrange in a firefight than speaking in front of people. Besides, what story can I tell?"

"DT." He laughed that I could ask that question. "What story could you tell? Sir, let's consider this. You got blown up in Afghanistan. You had a fifteen percent chance to survive at all. You almost died three times.

Eighty percent of your body has third-degree burns. You were in a coma for four months. When you woke up, they said that you'd be in the hospital for eighteen months, that you'd be on a respirator for the rest of your life. After that prognosis, you left that hospital walking and breathing on your own. And here you are now, affecting policy and making change. If that's not a story, I don't know what is."

I understood why Moseley invited me. He knew my story could resonate with service members and their families, to get them prepared for what could happen. God forbid it could happen to them, what happened to me, but it could. I could help them get ready for it, to see the possible outcomes of even the worst possible scenario.

"You got me there, sir," I said. "I won't tell *you* no anyway. You're the chief of staff of the Air Force. I won't say no to you."

That's how I began my career in public speaking. I started with military groups, then schools started calling me, then colleges, then charities, and then celebrities. Sports celebrities began inviting me to speak to professional teams, and owners of professional teams would have me speak to their men and women.

At first, I wasn't very good. I used a lot of colorful

words. I was just talking to the crowd like I was talking to my guys, and they wanted me to make it a little more family appropriate. My audiences and critics gave me consistent feedback: "Great speech, great story, but tone down the language a little bit."

I tried, man. I really tried.

In 2009, I was invited to be the keynote speaker at a USO event in Washington, DC. There were all kinds of Medal of Honor winners there, lots of recipients of military awards—and it was the biggest event I'd ever done. I was nervous, and I think my tablemates could tell. I think probably anyone could tell.

One of the people seated at my table was actor and comedian Jon Stewart, then the host of *The Daily Show.* Jon had entertained thousands of servicemen in USO events all over the world, and he struck up a conversation with me that night, asking my story. He asked if I felt okay about my keynote. I told him I was a little bit nervous, and he was like, "No, man. You got this. It'll be awesome."

I took the stage and delivered my speech, and I felt like I went blank for a minute on the stage, not really sure what I said. But when I finished, the whole place applauded. All those Medal of Honor winners rose to their feet, and I got to honor them.

I returned to the table, and Jon Stewart said, "Dude, I told you! You killed it!"

I sat down in my chair, feeling nothing but relief. I said, "Beer me, bro." Jon and I have been friends ever since.

I also became good friends with Bob Woodruff, whose story has a lot of parallels with mine. Bob was reporting for ABC's *World News Tonight* when his armored vehicle was hit by a roadside bomb. His world changed in an instant, and Bob spent thirty-six days in a medically induced coma. As Bob recovered from his injuries and his experience, he got to know the families of other service members dealing with similar injuries and experiences. He started the Bob Woodruff Foundation, whose mission is to make sure that veterans, service members, and their families have access to the highest levels of support and resources they deserve, for as long as they need them.

In 2007, the Woodruff Foundation partnered with the New York Comedy Festival to launch Stand Up for Heroes, a charity event featuring top comedians, musicians, entertainers, and A-list celebrities. Stand Up for Heroes has become known as one of New York's most anticipated nights of hope, healing, and laughter. Their event honors our nation's injured veterans and their families, and the Bob Woodruff Foundation has raised more than $55 million to create ongoing positive resources and lasting outcomes. I'm so honored to be part of it.

I've met some pretty amazing people at this annual concert. One year, we even met Richard Marx in person.

Jon Stewart often spoke or performed at Stand Up for Heroes, and our friendship grew as we kept showing up at those events over the course of more than a decade. Obviously, he's a liberal, well known for his views on *The Daily Show*. I'm a conservative, and we basically don't agree on any political policy. Lots of my family and friends ask me, "How the hell are you two so close? How can you possibly be friends?"

We're the odd couple, you could say. But he cares a lot for veterans, and so do I. And you can build a strong friendship when that's your common ground. Sure, we bust each other up on other issues, but it's mutual, never disrespectful. I've got his back, and he's got mine.

After numerous surgeries, skin grafts, grueling physical therapy, and coming to terms with our infertility, I had one more goal to accomplish. I wanted to get back in the Air Force. From the moment the doctors told me my military career was over, my whole goal had been to reenlist. I wanted to continue to serve.

When I talked about that dream, people would ask me, "DT, why? You have given enough. You're becoming good at other things—sports and speaking and teaching. Communicators make good money on the speaking circuit. Why not do that instead?"

But it wasn't about that. It wasn't about the money. There are thousands of people who make a lot of money and hate their jobs. I loved being an operator, being a TACP, being in the Air Force, and serving my country. I wanted to get back in.

People questioned me. They said, "But DT, you can't run anymore." So I showed them that I could.

They said, "But DT, you can't ruck anymore," the term for running a twelve-miler with a backpack. I showed them that I could.

I went through parachute training a second time, along with cadets. If it was at all possible, I was not going to miss my chance to jump out of a plane again.

It took about five years for me to attain the level to prove that I still had value, that I could still contribute to the Air Force. In December 2009, I had my medical evaluation board to ultimately determine if I could reenlist. My hope was in their hands.

Any airman can face a medical evaluation board at any point during his or her career, and this process is in place to ensure the safety and well-being of all the members of the Air Force. The board determines whether or not a member is fit for duty, and whether or not they will have any imposed restrictions.

The evaluation board consists of three physicians, and some cases require a psychiatrist as well for mental health conditions. The evaluation can result in one of two recommended actions: "return to duty," or "refer to informal physical evaluation board." I desperately hoped I had proved myself. I wanted that recommendation: return to duty.

It can take a couple of months to get the verdict, and that can feel like a long and stressful process. The fate of the airman's career is on the line.

The board called me to say they had made a decision, and they were ready to present their verdict. I walked in, along with my case manager. All the doctors were seated. Their faces were stern, and their voices were grave. "Sergeant Del Toro, here are our findings. We have determined you to be one hundred percent disabled, eligible for medical retirement."

Crap, I thought. I was devastated.

"That's not all, DT."

I held my breath, waiting for more.

They said, "We also have a second option: while we have determined you to be on one hundred percent disability, we have also decided you can come back as an active-duty instructor."

My heart soared back like a boomerang.

"Yes. Yes, sir. Where do I sign?" I asked.

"Well, one more thing," they said. "We also have a third option for you, DT. One hundred percent disability, you are eligible for military medical retirement, but you can come back as a GS-11."

GS-11 includes white-collar positions in the Air Force, which meant I would have had to retire and come back as a civilian. They invited me to be a government employee, as a civilian instructor, making a lot of money.

This was a curveball, and I needed to think it through.

I went home, and I called two of my mentors, Marty Klukus and Troy Lundquist, asking what they thought I should do. They each said the same thing: "DT, I don't know why you're calling me. You know what you want to do."

All through the years of my recovery, in all the interviews of people asking me why I wanted to reenlist, I said I wanted to join my team again. I wanted to keep serving. How could I have any honor if, now given the choice, I decided I'd get out and come back as a civilian?

They were right. I knew what I wanted to do, and it wasn't happening in an office. Sometimes you just have to hear someone tell you what you already know. I went to the Air Force the next day to tell them I had made my choice: I'm staying in.

In February 2010, I became the first 100 percent disabled airman ever to reenlist in the Air Force. I became an instructor, teaching the next generation of operators, to get them prepared.

I was a little disappointed that I couldn't deploy. Would I still love to be downrange as an operator? Sure, but I knew that couldn't happen yet. At least not until I got a cool robotic hand like Luke Skywalker. We're not there yet, but I believe in the people who are exploring this stuff every day. Maybe someday.

But I knew I could teach. I knew my mind was there. (My wife and my friends might disagree.) I threw myself into the job of training others for the dangerous and critical job of being a JTAC. With my combat experience, evidenced by the scars on my body, I had earned unequivocal credibility and unquestioned authority as an instructor. To prepare the next generation of operators for what is coming, that would be my greatest honor.

CHAPTER 22

All the while, I continued to explore adaptive sports, rising higher in the performance pipeline. All warriors can compete in the recreational sports, and then we can choose to rise in the games as we begin to excel. Beyond basic levels, there are developmental and emerging levels which become more competitive and have a broader scope to local and regional competitions. The highest levels of competition are the Warrior Games, and then the Invictus Games.

The Warrior Games launched in 2010, as a movement of Mr. John Beckett and a partnership with the US Army at the ESPN Wide World of Sports Complex. The games began with the intention to motivate service members to get out there and live again after an injury. The goal of Warrior Games was to get men and women into sports so they could have confidence again, to show the young people some older guys who made it through to the other side. It truly didn't matter if you won anything, just that you did it.

The founders of the Warrior Games came to me and said my resilience and influence made me a great fit. That made a big difference to me, to have these guys who came before me, to show me it could be done. By now, you know those words will always put me in the game.

At first, Warrior Games included only the United States military, service members of the Air Force, Navy, Marines, and Army. Then, a couple of years into the Warrior Games, the British service members piped up as well. In 2012, the British started competing in the Warrior Games, sending their wounded, injured, and ill to participate. That's how Prince Harry found out about it, and that's how I met Prince Harry.

We had an instant connection. He was a military guy, and we became friends. Prince Harry served for ten years in the British Army, and he was deployed to Afghanistan twice. He served as a forward air controller for the UK, and served as a TACP on his first tour of Afghanistan. He and I had held the same position.

My call sign was Gunslinger, and Prince Harry's call sign was Widow Six Seven.

He came to the Warrior Games and watched us play exhibition sitting volleyball, and then he joined me and beach volleyball Olympian Misty May-Treanor. Not too many people can say they got to spike a ball on a prince, but I can. That was a good day.

After the volleyball game, Prince Harry and I talked about all that we had in common, how we'd served in the same position. (There may have been some smack talk about who was the better TACP.)

Prince Harry saw the potential in the Warrior Games, and he wanted to expand it to involve all nations, all military servicemen worldwide. He's the prince of England, so he could do that if he wanted to. Prince Harry started the Invictus Games, a world-class competition for the world's most athletic wounded warriors. Only one hundred were chosen to represent Team USA, and I was honored to be among them. In 2014, I competed in the Invictus Games for the first time.

Meanwhile, in 2015, the Air Force sent me to Colorado Springs to train as a para world-class athlete. I started training at the Olympic Shooting Center, the largest indoor shooting facility in the western hemisphere, and the third largest in the world. I had years of training as an expert marksman, and this was my jam.

I was invited to go to the US Paralympics Track and Field Championships to compete in shot put, discus, and javelin, and I set all new world records that day in California. I started to win medals. I began breaking world records for my class. I set world records in track, and I was also picked up to be part of the USA shooting team with my air rifle skills. Currently, I hold two world records for shot put and javelin. In 2014, I won a silver medal for powerlifting. In 2016, I won the gold medal for shot put.

When I think back to the beginning, I started adaptive sports because I was trying to get back in shape, and I just wanted to find my place in society. But then I started winning. I never intended to become the best, but then, I actually did become the best. Who would have thought?

As veterans, there is a lot of focus on the guys who died, as there certainly should be. They made the ultimate sacrifice. But there are guys who came home, wounded, with lives left to live. I wanted to represent them in the games. I didn't expect to win anything in those competitions, but I saw them as an opportunity to show the world—this had sucked, yes, but I'm still living my life, *and enjoying it.*

Let me tell you: there's no feeling in the world like thinking you can't do something anymore, and then discovering that you can. I didn't care about the medals or chase after them; just like in the Air Force, I was there to do my job, not win medals.

But I'm not going to lie. The medals were pretty cool.

In 2016, Prince Harry announced that the Invictus Games competition would be held in the United States. It was a worldwide event, hosted at Disney's ESPN Wide World of Sports Complex in Orlando. Competition was rising all over the world, and President Obama and First Lady Michelle Obama even launched a social media video challenge to Prince Harry. Their video of smack talk made the rounds on YouTube.

"Hey, Prince Harry," POTUS and FLOTUS said, "remember when you told us to 'bring it' at the Invictus Games? Be careful what you wish for."

Not to be outdone, Prince Harry replied with a challenge from his grandmother, the Queen of England, throwing shade right back. It was a mic drop heard round the world, a fantastic use of influence.

When Michelle Obama and Prince Harry were interviewed by Robin Roberts about the Invictus Games, they each spoke about their passion for the military. Mrs. Obama admitted to the playful sparring with Prince Harry over which country would reign supreme. She wanted to see people cry tears of inspiration, to see warriors who would not give up. Each country wanted to bring home the gold, and the warriors would deliver. She said, "We've done some digging at each other over the past several weeks. But what do I really want at the Invictus Games? I want people to be inspired."

Meanwhile, Prince Harry's assistant called my cell phone. He said, "Prince Harry would really like for you to be the speaker at the opening ceremonies." That was quite an invite! Again, he's the prince of England, so he can ask for what he wants, and I'll be there.

Saying yes to speaking at the opening ceremonies meant I was both speaking *and* competing. So, in the days before the opening ceremonies at the Invictus Games, we had days of rehearsals. Those were some very taxing days of practicing and training, of long hours in high heat.

Now remember, I was prone to some colorful language. As I rehearsed, I thought I'd end this speech the same way I ended every speech: "Stay strong, finish strong, and never fucking quit."

But at the Invictus Games, they didn't want me to swear. They wanted me to say "never, ever quit." It's a reasonable request, even if it doesn't pack the same punch. But every time I came to that line, I kept forgetting to replace it. I kept stumbling over it, saying the word they didn't want me to say.

I promise I was trying to filter myself. Believe me, I knew what a big deal this was. The Invictus Games is one of the biggest live simulcasts, and I had to get this right. But if you tell a person not to say a specific word, it seems to become the only word they can say.

Little did I know, Prince Harry was there, hiding away in some control room, watching over the rehearsals and hearing every mistake.

Later that afternoon, I attended a wheelchair rugby exhibition game. Prince Harry came up to me, shaking my hand. He said, "Dude, you're so funny."

"What? What are you talking about?"

"I've seen you," he said. "I've seen you at your rehearsals, and I see how you want to swear every time, but you hold it in. It makes me laugh every time."

My parents on their wedding day.
Photo by Ken Kraph.

My dad died when I was thirteen, and my mother died a year later when I was fourteen. She is buried in the cemetery with the fountain, just as she asked. This photo was taken the first time my son visited his grandmother's grave site. He was ten years old.

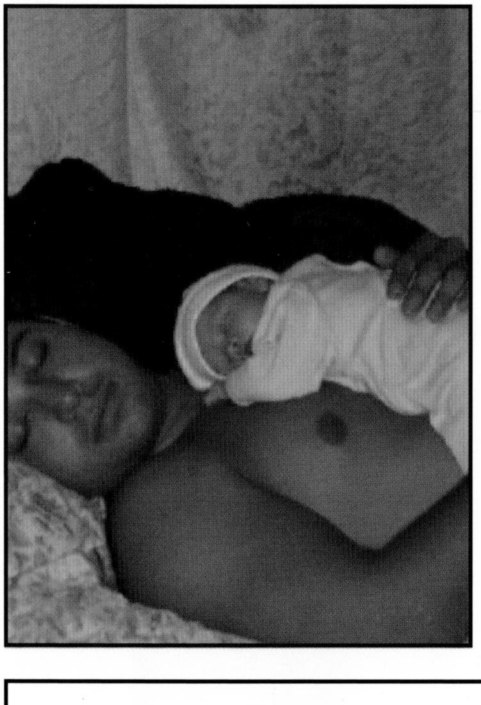

My baby boy and me. Carmen set this as her screensaver for the months of my coma, hoping that I'd be able to hold him again.

By October 1999, I was on a great high and living my best life. I was certified to call in the air strikes, and I became a fully qualified JTAC (joint terminal attack controller). I served in special operations—jumping out of planes, calling air strikes, and directing combat aircraft engaged in close air support.

This photo was taken days before the explosion, on my last operation before I was injured. That's a good-looking dude right there.

In the Air Force, men and women are trained to live and die with one another, for one another. While I was away from my family, these people became my brothers and sisters.

In every single way, I was living the dream. In the United States Air Force I was trained, prepared, and ready to defend this nation.

On December 5, 2005, this Humvee rolled over a buried roadside land mine. I was in the front passenger seat.

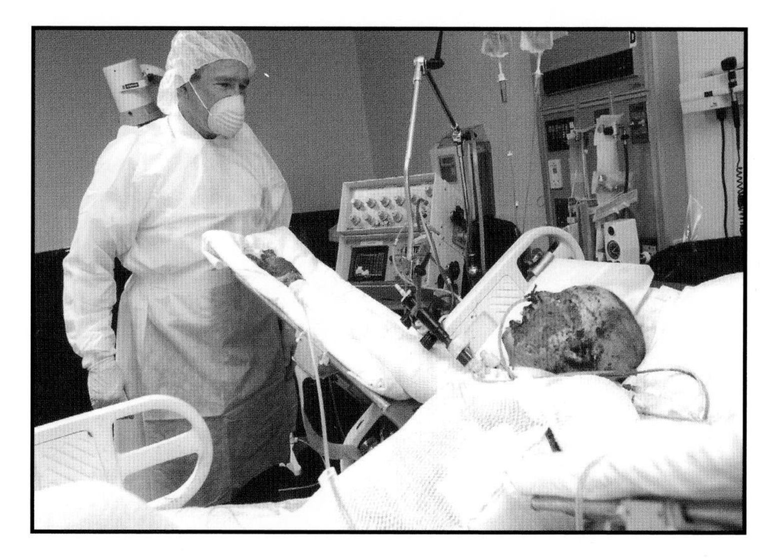

President George W. Bush visited my bedside in 2006, awarding the Purple Heart to me. I sure wish I could remember that.

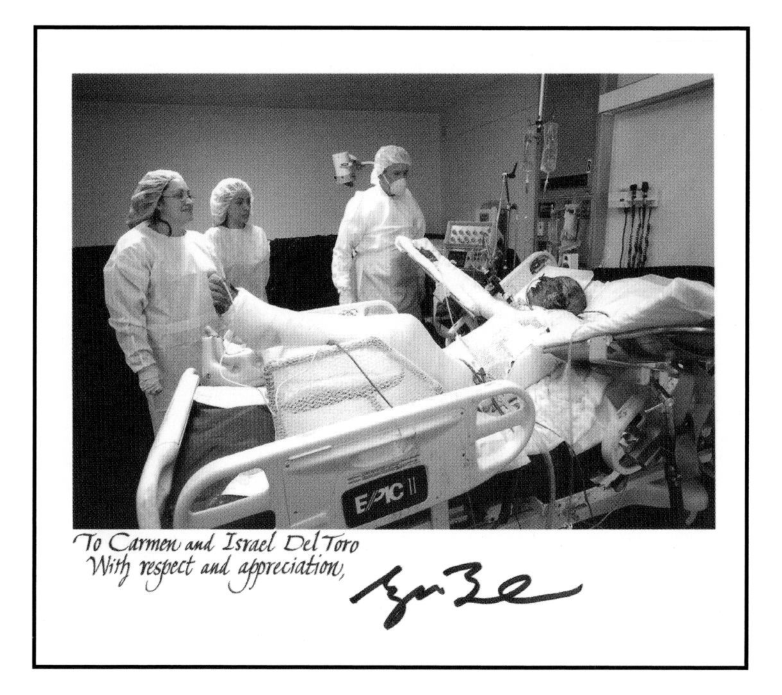

President George W. Bush sent me this signed photo of his visit to Brooke Army Medical Center on the day I received the Purple Heart.

My first adaptive sport was sitting volleyball. This became an important part of my recovery, helping me to restore my range of motion and develop dexterity with my hands. With this entry into society, I began to see what my life could look like after my injuries.

I played sitting volleyball against Prince Harry. Not too many people can say they got to spike a ball on a prince, but I can.

I trained in the throwing events, including shot put, discus, and javelin. With each additional sport, I discovered a part of myself that still lived inside me and was stronger than ever.

I also learned to excel in adaptive cycling, where the para athletes compete on hand cycles, recumbent bicycles, and tandems.

I biked the MS 150, a 150-mile bike race from Houston to Austin, Texas, in support of the National Multiple Sclerosis Society.

This is my good buddy Prince Harry and me, hanging out at the Invictus Games in London.

Here I am competing in one of my first shooting competitions as a para athlete.

Finally, I gave my bride the wedding I had promised her. We had a giant wedding reception with my teammates and family members from every part of our family tree.

President Bush has become a friend of mine. In 2017 he included me in his book, *Portraits of Courage: A Commander in Chief's Tribute to America's Warriors*. He included sixty-six stories and full-color portraits of veterans, each one painted by his hand.

Jon Stewart has become a good friend of mine. He presented me with the Pat Tillman Award on behalf of ESPN, and I presented him with the 2022 Mark Twain Prize for American Humor at the Kennedy Center.

I didn't go into this career for medals but, I'm not going to lie, they're pretty cool to have.

Even those of us who are injured cannot imagine what the journey is like for the caregivers. This woman is a hero.

My wife, Gueris, and my son, Guero.

At my retirement in 2019, I surprised my wife with a renewal of our vows. To finish the ceremony, my students created an arch with their swords, an Air Force tradition for newlyweds that marked the end of my career.

I have had the honor of meeting many presidents, including President Obama, President Trump, and President Biden, back when he was vice president of the United States.

Steve Carell and I honored Jon Stewart as he received the Mark Twain Prize for American Humor, recognizing his lifetime achievement in comedy at the Kennedy Center for Performing Arts.

This is my dad, the original Israel Del Toro. I have lived all of my days to keep my promises to this man, the hero of my life.

Busted. I said, "Sir, I apologize. I'm trying so hard not to say it."

He got that Prince Harry twinkle in his eye, and he said, "Just say it. Just do it."

"Are you sure, sir?"

A lot of the athletes had wanted me to say it, but I didn't want to offend anybody. I was working so hard to edit myself. But suddenly, Prince Harry asked me to say it. And he's the prince of England.

I said, "Sir, do you think you should check with your grandmother?"

He laughed, and he said, "Just do it, DT."

I gave it some thought. I knew it was a live simulcast, and I imagined there was a delay. If I chose to say it, I thought they might bleep it out. Was it a risk worth taking?

That evening, at the opening ceremonies, Prince Harry took the stage in Orlando. He said, "I cannot tell you how proud and excited I am to open the second Invictus Games, here in America. I'm a long way from London tonight, but when I look out and I see so many familiar faces—servicemen and -women, their friends, and their families, and all of the people who have gotten them here—I feel like I'm at home."

He talked about his own time in the military, how

he saw firsthand the sacrifices that the soldiers and their families made. He had joined the armed forces to be one of the guys—as so many of us do, but then he traveled back from the battlefield on a plane carrying the body of one Danish soldier and three young British soldiers, each fighting for their lives. He said that was when he began to understand the very real and permanent cost of war.

He said, "I saw the sacrifices you and your families made to serve your nations. I learned about the importance of teamwork and camaraderie, in a way that only military service can teach you."

The crowd applauded. He was speaking our language, talking right to our hearts.

He recognized that his privilege as a prince gave him the influence and ability to help his military family in powerful ways. Prince Harry said, "And that is why I had to create the Invictus Games. To build a platform for all those who have served. To prove to the world what they have to offer."

Prince Harry explained to the audience what they would see in the days to come: people who should have died on the battlefield, and now they would be going for the gold medal. He said, "Mark my words.

You will be inspired, you will be moved, and I promise, you will be entertained." (Given our private conversations, I felt right then like he was talking straight to me.)

And then Prince Harry took a unique opportunity to talk about an issue that desperately needed to be discussed: not just the physical injuries, but also the emotional and mental challenges of war. The long-term invisible wounds of PTSD.

As he paced the stage, he spoke into the microphone and said, "When we give a standing ovation to the competitor with missing limbs, let's also cheer our hearts out for the man who overcame anxiety so severe he couldn't even leave his house. Let's cheer for the woman who fought through post-traumatic stress—"

The crowded interrupted him with cheering and applause. He let us cheer, and then he began his sentence again.

He said, "Let's cheer for the woman who fought through post-traumatic stress, and let's celebrate the soldier who was brave enough to get help for his depression."

He spoke to the audience watching the simulcast from their homes, and he said, "Over the next four days, you will get to know these amazing competitors. They weren't too tough to admit that they struggle with mental health, and they weren't too tough to get help they needed. To all those of you watching at home, who are suffering with mental illness in silence—whether a veteran or a civilian, a mum or a dad, a teenager, or a grandparent—I hope you see the bravery of our Invictus champions who have confronted invisible injuries. I hope you're inspired to ask for the help that you need."

We raised our hands in a roaring ovation. Prince Harry spoke words of truth, words that the whole world needed to hear.

"To end, can I just say thank you to all of you guys. You are fierce competitors. You are role models that any parent would be proud to have their children follow. You've made me a better person, and you're about to inspire the world, and I'm proud to call you my friends. So, let's put on a hell of a show, in memory of our fallen comrades who didn't make it. We are Invictus."

He was flawless, that prince of England. He had brought the Invictus Games to Orlando for a worldwide event, and he gave us all so much to be proud of.

Next, the announcer welcomed the First Lady of

the United States, Michelle Obama. She came on behalf of Joining Forces, a nationwide initiative calling all Americans to rally around service members, veterans, and their families, to support them through wellness, education, and employment opportunities. She brought a whole different energy of her own, calling out into the crowd through her microphone, "Hello, everyone! Are we having fun tonight?"

She teased that she had asked the USA team not to win too much. She winked and said it's the least we could do, naturally, as hosts of the games.

She welcomed the devoted advocate for men and women in uniform, the honorary chair of the Invictus Games, former president George W. Bush, and former first lady Laura Bush. She thanked Ken Fisher and his family for their tremendous support of the event, and for their tireless commitment to our military community.

And then Mrs. Obama said, "Most of all, I'm here to honor all of you, our extraordinary service members, our veterans, and *of course,* our military families. You all are amazing. Truly amazing. I am so incredibly inspired by all of you. I'm inspired by your courage, by your love of country. I am inspired by the sacrifices you all make every single day—particularly the wounded warriors and the caregivers, who show such strength and resilience in the face of challenges that most of us can't even imagine.

"In the course of your recoveries, you've endured surgery after surgery, and months—even years—of painstaking rehabilitation. And you *still* tell us that you'd do it all again, if you were given the chance to serve your country."

As a crowd, we thundered in agreement. Absolutely, we would do it all again.

She said, "And look at you now! Today, you are here competing in these games. You are achieving amazing athletic feats! I am so proud of you all! And as for our caregivers—let's give it up for our caregivers!" And the place roared once again, as we cheered for those we couldn't do without.

She said, "We know how you all dropped everything when your loved one was injured. How you stepped up to manage all of the appointments, the medication, all the finances, often juggling a job and raising kids at the same time."

Amen, Mrs. Obama. Even those of us who are injured cannot imagine what the journey is like for the caregivers.

She said, "Ultimately, that's really what the Invictus Games and Joining Forces are all about. They are about shining this big bright light on all of you—*all of you*. They are about sharing your stories and your achievements with the world, so that we can all be inspired to live up to your example.

"That is why we're all here tonight," she said. "That's why I'm here tonight. And that's why I'm so proud of you all. That is why I love you all so much. That's why I will keep working for you all, as long as I'm alive."

"I wish you all the best of luck in your competitions," Mrs. Obama said, "and I hope you have a lot of fun along the way."

And then, she introduced an ambassador of the Invictus Games, a legend of the silver screen and an advocate for warriors, Mr. Morgan Freeman. He walked onstage in a sleek white suit, looking like the classy guy that he is.

He said, "Tonight, we are privileged to hear from wounded servicemen and -women, as they share their journey from darkness into light. We will hear how the power of sports transformed their recovery, self-confidence, and joy of life. These men and women have been through hell, and they have survived against all odds. Sometimes it is easy to forget that being wounded in service, or taken ill while enlisted, affects more than just the individual. The stories you will hear embody the true meaning of Invictus."

Then he said, "It gives me great pleasure to introduce one of the faces of the games, Master Sergeant Israel Del Toro, Jr." He turned to face me, and he cheered my name, "DT!"

Let me tell you, in my life before that day, never once did I ever imagine Morgan Freeman would cheer my name.

I shook his hand and took my turn at the stage. I spoke to them about strength, faith, and honor, and doing what it takes to take care of our families. What an honor to be in the presence of so many warriors and their families, each one a hero. The audience was filled with warriors who understood that kind of conviction. They stood to give me an ovation, and in my heart, I was standing for them, as well.

The music swelled in the background, and the crowd roared with applause.

When I told the story of Guero, my spark and my life's inspiration, Guero came out onto the stage, wearing his dress shirt untucked, a vest, and a tie. My beautiful wife joined me on the stage, and she came to my side. The whole place cheered on and on for my family. I motioned for them to hold their applause, but they wouldn't stop.

So, I returned the applause right back to the warriors and their families.

I said, "That is for anyone who has been injured. I beat the odds. And, like so many others, it was only possible because I had one group of people who rallied around me and supported me. My teammates on the field, my medical staff, the surgeons, the therapists, and the love of my amazing wife and son."

I looked to Guero and Carmen as I said those words, and Guero saluted me.

"Because of this support, I've been able to conquer all these obstacles. Like the legendary phoenix, I am reborn from these ashes. These flames have made me stronger."

On the giant screen behind me, there appeared the image of a phoenix, born from the ashes indeed. The crowd cheered, and I knew, this was my chance. I spoke into the crowd, and I gave them every drop of passion I had.

Spurred on by Prince Harry, I said, "Stay strong. Finish strong. And never fucking quit!"

And the whole stadium erupted. I looked at Prince

Harry, and he was loving it. The place went crazy! They didn't want me to be censored. They wanted the truth, and I gave it to them.

And as the crowd roared on, someone said, "We have a surprise for you. Take a look."

I turned behind me to see President George W. Bush standing there with my family. We met again.

My commander in chief saluted me, and I saluted him in return. He put his arm around me, he turned and gave Carmen a kiss on the top of her head, and then he gave Guero a high five and a handshake. He showed his support to my most important people, there in front of tens of thousands of people.

Away from the microphone, I realized what I had said near my commander in chief. I said, "I'm sorry, sir."

He smiled his famous grin, and he said only to me, "Oh, DT. It was great."

Into the microphone, he said, "I'm really proud of you, my friend. And I want to thank all of those who serve our country. I love your example. I love the fact that you made it. And I love that when they told you you'd never walk again, you said . . . well, I don't want to repeat the word."

This time, the audience laughed. They knew what word he meant.

President George W. Bush said, "I want our fellow citizens to know that if you think your life is miserable, look at my pal DT. He's overcome a lot. And he's a huge contributor to our country, and I'm really proud to call him friend."

Then he turned to the cameras and spoke to the audience at home. He said, "For those of you who are out there watching, who may have been in the military, if you've got an invisible wound of war, follow his advice. Seek help and contribute to the future of our country."

President George W. Bush turned to me, and he said, "God bless you, my friend."

And then he turned to the crowd of warriors, and he said, "And God bless America."

Our former president shook my hand once more, and he hugged my wife once again, and we followed him off the stage as the crowd chanted, "U-S-A! U-S-A!"

It was epically amazing.

Morgan Freeman returned to the stage, and he said, "This story touches our hearts today. But it's all the more powerful when you realize that the story of this one family is told in thousands of similar stories around the world."

Yes, Mr. Freeman, that will always be true.

I am one of many, and I stand on their behalf.

CHAPTER 23

One afternoon in 2016, I was working at the Wings of Blue, the Air Force Academy's parachuting course, when I got a phone call from someone at ESPN. They casually introduced themselves, like this is a phone call I get every day. And then they said, "Sir, we would like to invite you to the ESPYs."

Listen, I know the ESPYs. It's the only award show I watch. I don't care about all the other stuff. You can have your Oscars and your Grammys all day long, but the ESPYs is the one award show that I do not miss.

I knew a couple of their sports reporters from the Invictus Games, so I was thrilled to be invited. I hung up the phone, all excited, high-fiving with my workout buddies, telling them, "Dude, I just got invited to the ESPYs."

Then ESPN called me right back again. "Sergeant Del Toro? It's us again. Listen, we failed to mention something on our last call. Sir, you're in the running for an ESPY, for the Pat Tillman Award for Service."

"What? No way. You're messing with me."

"We're serious, sir. The foundation will contact you soon with details, but we wanted to get in touch with you to make sure you're in attendance that evening."

"Well, thank you. I will be there."

I knew who Pat Tillman was, and I felt honored just to be mentioned in the same sentence with him. That man is a legend. Pat Tillman played in the NFL, and he left his sports career to enlist in the Army in the aftermath of the attacks on September 11. He served several tours in combat before he was killed in action as a result of friendly fire. Since 2014, the Pat Tillman Foundation has been issuing an award each year in honor of his legacy.

I hung up and told my workout buddies the update. "Dude, you won't believe this. I'm *in the running* for an ESPY."

"Shut up. You're such a liar."

Frankly, it was a fair response. The whole thing

was unbelievable. It took some convincing, but I told them I wasn't joking. This was really, actually happening. The ESPYs. Pat Tillman. Me.

A day or two later, the director of the foundation called to confirm. She said, "Yes, sir, this is happening. We can't announce who wins until later . . . but honestly, I hope you win."

A couple of weeks passed, and they called me to tell me I had won.

Now I was the one in disbelief. I said, "Shut up. You're lying. Are you serious?"

"Yes, sir," she said. "Congratulations. You'll receive the ESPY live on the show."

I hung up the phone thinking, Holy crap, holy crap, holy crap. It was the biggest news of my life, and I couldn't say anything to my buddies. I couldn't say anything to anyone. I had to wait until they announced it.

I was competing at the Warrior Games when the news was officially released. The Warrior Games announced, "We want to let everyone know that one of our own athletes is about to receive the ESPY for Pat Tillman. This year, the Pat Tillman Award for Service will go to our very own Israel Del Toro, Jr."

The place went crazy. It all felt like a dream.

ESPN's writers wrote a speech for me to deliver at the ceremony, but they wanted me to talk about my sports achievements and my medals. Honestly, it sounded nothing like me. No thanks, I really didn't want to talk about any of that. I met with the speechwriter, and I said, "Listen, dude. Can we rewrite this?"

We rewrote it within an hour. I didn't want to say any of that shit about myself. I had other people to talk about. I got to say what I really wanted to say: a giant thank-you to the people who saved my life.

On the night of the ceremony, none other than my good friend Jon Stewart presented the award to me. With the Pat Tillman award in my hand, I thanked his widow, Marie Tillman, and the Tillman Foundation. I recognized that Pat was a prime example of service before self, and knowing that he was an operator, just like myself, meant even more to me and my SOCOM community.

I said, "Receiving this award is still strange for me. I don't see myself as someone special. I just did what any other service member would do. I tried to make things better for the guys that follow me, and I tried to take care of my teammates. That's what we do."

The cameramen focused in on my son in the audience. His handsome face filled the screen. "Guero, you're my fire. You're my spirit. You are the one who keeps me pushing every single day. You are my greatest accomplishment. To my wife, you are a pillar of strength. You are an amazing and outstanding woman, to be able to handle what you did, by yourself and with a three-year-old son. And to my dad, hopefully I'm living up to the last words you said to me: 'Always take care of your family.'

"So, to all my wounded, injured, ill service members, disabled civilians, or anyone who's having a bad day, I will stay strong for the ones who cannot. I will fight for the ones who cannot fight. And I will never fucking quit on you. I will finish strong. This is my promise, this is my pledge. Thank you, for letting this guy—somebody who just had a bad day at work—feel like someone special tonight."

CHAPTER 24

All military personnel are subject to continuous evaluation, and anyone who has a secret or top secret clearance is subject to this ongoing screening process to maintain their eligibility. Every time I wanted to continue my career, I had to enter the same process, do all the paperwork, and prove myself again. I was getting kind of tired of doing that stuff. In November 2018, I started thinking about my retirement.

I went back and forth: to stick around or to retire. A lot of people kept trying to talk me out of it. I knew I was up for a promotion. I would have made rank as chief master sergeant, E9. I knew if I made that, I would have to move wherever the Air Force wanted to send me. By then, Guero was going to be a senior in high school, and I didn't want to move our family again, especially not during his last year of high school. I had been stationed in North Carolina, Florida, Georgia, Italy, Korea, San Antonio, and finally Colorado Springs. I began to think it might be time to call this place home.

General Goldfein, chief of staff of the Air Force, called me on my cell phone to ask me to change my mind and stay. That's some serious pressure right there, listening to the chief of staff of the Air Force say, "DT, please stay. We need you. You're going to make chief."

"Sir," I said, "I really appreciate you calling me and trying to talk me into staying. But if I'm staying just to make rank, then it's not for the right reason. I appreciate your confidence in me, and it means everything that you'd like me to stay. But my family sacrificed twenty-two years for me, and it's time for me to sacrifice for them. It's time for me to wrap this up."

When I finally made my decision, I told my commander it was time to pull the trigger. I wanted to involve my students and my teammates in my retirement, and I wanted to retire on my birthday that year, which happened to fall on a Friday. I made my decisions, and we started all the planning.

Carmen and I had been to dozens of military ceremonies, but a retirement ceremony is different. It marks the end of a long career, as well as the transition into civilian life. It's a big freaking deal. I decided I wanted to do one last leap out of a plane. I started with a jump in my career, and after twenty-two years in the military, I wanted to end with a jump.

As you can imagine, my Tía Gris did not want me to jump. She was there in the earliest days of my injury, and she was not interested in anything going wrong. Tía Gris said, "Haven't you had enough? Why do you have to keep doing this?"

But I couldn't explain it. Jumping is in my blood now.

Early on the morning of my last day, I made my way to the runway for one last jump with the Wings of Blue. I jumped with the American flag, so anyone nearby could see this was no ordinary parachute jump on an early Friday morning. This was a celebration, a finish line, a salute in the sky.

I landed safely, and I unzipped my jumpsuit to show the T-shirt underneath. It read, THE LEGEND HAS RETIRED.

We held the ceremony at the Air Force Academy, and it was beautiful. Patricia Vonne sang the national anthem, my commanding officer gave a speech, and two of my friends gave speeches as well. There was a parade of colors, and the Old Glory presentation. My teammates lined up in order of ascending rank, and I was the last in line. They passed the flag from the lowest rank on up, a representation of my progress through my military career.

I presented gifts to my family: to Carmen, I gave a bouquet of roses. To Guero, I gave the knife I had carried for all of my military career. I had carried it with me everywhere, through all my deployments. For this day, I had it engraved with a message just for my son: *Guero, you are my spark.*

At the end of the ceremony, I had one last surprise. I said, "One more thing. To show my dedication to my wife, I'd like to renew my vows."

Carmen had made it tricky to pull that one off, but her sisters helped me out. They made sure her parents were in attendance, and they encouraged her to wear a white dress. She couldn't understand why they so badly wanted her to wear white, and she really wasn't having it. Leave it to the older sisters to make that magic happen. It all came together.

In that moment, the priest came out, my cadets lined up, and I renewed my vows to my wife, before God and everyone. After all she had given me, I wanted to honor her with these promises as I finished my career.

My students created an arch with their swords. This is a tradition for newlyweds, an archway for a bride and groom to walk through, and quite fitting for the renewal of our vows. We walked through the swords to leave the chapel, but the last two guys lowered their swords down. They wouldn't let us through.

"Sir, give her a kiss." I gave her a peck.

"Sir, that isn't good enough! Give her a *kiss*!" So I laid one on her.

They opened up the swords to let us through, and then they whacked us both on the behind as we walked through. (Carmen was not expecting that.)

A friend of ours had a big barn, and he shared it with us for a giant party after the ceremony. We had tacos, a DJ, a mariachi band, and a bartender, because I knew we needed someone to temper the flow of drinks, or my teammates would throw them down before the night even began. As a final celebration to share with our guests—and a fancy gesture at that—we opened a retirement gift from Jon Stewart: an entire case of Dom Pérignon.

Together we raised a glass, to a career now finished. Cheers.

CHAPTER 25

Imagine this with me. Think about when you woke up this morning, yawning and stretching, taking the first deep breath of the day. One that fills up your lungs and rushes oxygen to wake up your body. (It's been a lot of years, but I still remember deep breaths like that.)

Now, imagine putting a narrow plastic straw to your lips, and take that same deep inhale and exhale. The only air you can take in is what you can get through that small opening of the straw. It's not quite as fulfilling, satisfying, or oxygenating.

Now, with the straw still in your mouth, put on a girdle. Cinch it up as tight as you can around your middle—and I mean *tight*. Close it in around your ribs, so every organ inside you feels tight.

Then, with the straw in your mouth and the girdle around your middle, climb up a flight of stairs. Maybe two flights of stairs.

Now, live at high altitude, where even healthy people are short of breath because of the oxygen levels.

If you're imagining your lungs feeling tight, hungry for more air, then you're close to what it feels like to live with the effects of inhalation wounds. All day, every day.

Now, with your girdle in place, your straw in your mouth, your stair climbing, and breathing at high altitude, imagine someone asking you to put on a mask. Imagine them requiring you to wear a mask to leave your home.

That's what happened in 2020, when the coronavirus swept through the world and restrictions began.

Covid was hard on our family in a number of ways. For Carmen, the scents of the pandemic brought back so many hard memories for her. Scent is the most powerful trigger for memory, and hand soap brings back the very worst of her memories. The protective gear for hospital staff, the smell of the soap—it's all familiar, and it takes her back to the worst time of her life.

On a global scale, the Centers for Disease Control and Prevention issued mandates all throughout the world. They considered the needs of Americans with disabilities when they released the mandates, and the CDC even quoted the ADA in their statutes:

> For people who cannot wear or safely wear a mask because of a disability as defined by the Americans with Disabilities Act (ADA) (42 U.S.C. 12101 et seq.), consider reasonable accommodation for workers who are not fully vaccinated, who are unable to wear a mask, or who have difficulty wearing certain types of masks because of a disability.

It's right there. The Centers for Disease Control included an exception for us in their mandates, but people didn't read that part. I guess they didn't hear that part, either. The words of the CDC seemed to override the laws of the ADA, the Americans with Disabilities Act. People seemed to ignore that fine print.

They were taking their instructions from media headlines and local leaders, without recognizing that some people could not abide by those rules. Not for political reasons, not because we wanted to take a stand against authority . . . but because we physically could not. I could not. And I was not the only one.

It would be convenient if everyone with a disability could fit neatly into one box, all of us with the same injuries, illnesses, wounds, and needs. But that's not how it is. A guy with burns has different needs from a guy with a prosthetic leg. And even a guy with a prosthetic leg below the knee has different needs from a guy who has a prosthetic above the knee. We don't all have the same list of needs.

I watched the headlines, I saw the tensions rising, and I knew this wasn't going to go well for people with disabilities. I watched to see how community leaders and lawmakers would represent those of us with disabilities. I waited to hear them speak for us, protect us, and speak on our behalf.

But they didn't. They didn't say a word.

Politicians and local leaders handed out mandates and rules, stating that a local business can refuse you service if you're not wearing a mask, just like if you were not wearing a shirt or shoes.

Listen, I know how to comply. You don't rise through the ranks in the military without knowing how to follow the rules. I can put on a shirt, and I can put on shoes.

But I cannot heal my inhalation burns, so I cannot wear a mask over my face.

"But DT, you should wear the shield," people said, and I wish I could. My skin has been grafted all across my face, and the nerve endings are different from yours. I cannot wear the shield because the foam against my forehead makes me overheat. I cannot bear the irritation.

"But DT, you should wear the mask across your nose, one that loops behind your ears," people said, and I wish I could. But my ears burned off. I do not have ears.

When the mask mandate went into place, I immediately asked for a letter from my doctor, stating I could not—and should not—wear a mask, based on my medical condition. I carried that letter as a file on my phone, showing it to people anytime I needed to.

Early in the pandemic, I stopped by my bank to deposit a check. A line of people stretched down the sidewalk outside the entrance, each person standing six feet behind the person in front of them. I took my place in line, minding the gaps of physical distancing.

When I got to the front of the line, the greeter at the door said, "Sir, do you have a mask you can wear?"

"I cannot wear a mask," I said.

"Then you cannot come in," she said.

"What do you mean I can't come in? I have a medical condition that prevents me from breathing when I wear a mask."

"I'm sorry, sir, this is the mandate by the state governor."

"There is also a mandate by the Americans with Disabilities Act that says you cannot require this of someone who cannot do it."

She said, "If you prefer not to wear a mask, you can go through the drive-thru to process your transaction."

"No, I cannot handle those."

"You can't *handle* them?"

She looked at me like I was being a smart-ass, a sassy punk who "absolutely cannot deal" with a situation.

I said, "Ma'am, do you see my hands?" I held them up, showing her the partial fingers I have on my disfigured hands. "I cannot maneuver the cap of the cylinder canister. I have an issue sending the money through the tube."

"Sorry, sir. I cannot let you in."

Here's the deal. I have a great bank, and they have a great staff. Most of the time, they have served me well. To this day, I am surprised they couldn't find a solution, but on that day, they didn't solve the problem. They placed a person on the front lines of entry, they gave her a job to do, and they told her a script to repeat. She was just doing what she was told to do, but she had been misinformed. I left the bank with my money still in my pocket, instead of deposited into my account.

Our local news crew is a team of talented people, and many of them are good friends of mine. Dianne Derby is an Emmy award–winning coanchor for the evening news, and by then we had done a few stories together. I gave Dianne a call on my way home, and she followed up immediately with a story that released the next day: "Veteran Denied Access to Credit Union for Not Wearing a Mask."

As Dianne interviewed me for the news segment, I received a call from the bank's vice president of communication. She knew who I was, and she was nearly in tears as she apologized on the phone with me.

I said, "Ma'am, it's okay. That person at the door, she was just doing as she was told. I'm not trying to get her fired, and I'm not going to sue you. You guys have had a hard year already, with the quarantines and shutdowns, so I'm not going to cause more harm. It was a brain fart. You guys just didn't think it through all the way. Look, 99.999 percent of the time, I haven't had an issue with you guys. But I do want to help inform you of the laws in place to protect people with disabilities. Yes, there is a mask mandate in place. And there are exceptions in place as well."

A similar incident happened at the commissary, the on-base grocery store. Because the commissary is one of the exclusive benefits of serving in the United States military, you must have your ID card with you to buy your items. As long as you have your ID, you're good to go. Carmen needed some fresh ingredients for a salsa she was making, and she asked me to stop by the commissary before I came home.

There was a young airman at the entrance, and we went through that same routine as the lady at the bank. He asked me where my mask was, and I told him I do not wear one, I cannot wear one, and I have a letter excusing me from the mandate. I showed him my military ID, and I offered to show him the letter from my doctor.

"Sir, I cannot let you in," he said.

"Are you telling me I cannot get into my own commissary, because I cannot wear a mask due to the injuries I sustained from serving my country?"

"Yes, sir. I'm sorry."

I shook my head, turned away, and walked to my car. Once again, I called Dianne. Her team ran another story: "Veteran Denied Entrance Again." As her team researched the story, they received a public apology to me from the colonel.

Colonel Johnson said, "I am grateful he brought to our attention that our Covid safety policies did not account for medical concerns of our veteran, wounded warrior, retiree, and family member populations. We must and will do better to ensure similar situations do not occur."

Thank you, Colonel Johnson. That's all I'm asking for.

Again, I was not asking for anyone to get fired. But some things need to change. I prefer not to say that I'm trying to *educate,* because I think that word is overused and overplayed. Educate? Not necessarily. Inform? Absolutely.

The commander of the base called in response to the story, inviting me to evaluate the places where they could improve their services to the most vulnerable and disabled. This, I could do.

You remember those months of quarantine—we all do. It was the first time any of us had experienced a global pandemic. For those couple of years of quarantines, vaccinations, boosters, and fluctuating mask mandates, I lived it right alongside everybody else, with the added responsibility to use my voice for the people who couldn't.

In November 2021, I was in New York City once again for the red carpet event for Stand Up for Heroes, their fifteenth anniversary. It had been months since I had to worry about masks, as people were finally becoming more aware of the needs of those of us with disabilities, but we still had a long way to go.

The morning of the event, I got a call from the venue—Alice Tully Hall—saying that I could not come to the red carpet event unless I had completed a rapid Covid test. I was supposed to have one the day before, but I hadn't received that notification. Nobody told me I had to have it done, so I missed that window of opportunity to get the negative test result. So, no red carpet for me.

I was disappointed to miss that part of the event, but that's fine. I can take responsibility for missing that piece of information. I'd show up to the main event, no problem.

They finished the phone call by saying cheerfully, "Okay, sir, we'll see you at the event. Don't forget your mask!"

Here we go, I thought.

"Well, I won't be wearing a mask. I cannot wear a mask."

There was a pause on the phone line, and the employee said they needed to check with some people about how to proceed. They hung up, and ten minutes later, they called me back with the decision.

"DT, we understand you will not be wearing a mask this evening, and so we need to let you know that you cannot attend the event."

"I have a doctor's letter that allows me to attend events without a mask. I'll send you the law that says I am exempt from this rule because of my disability."

Pause.

"Can you wear the shield?" I know what kind they mean, the face shield welders wear, only flimsier.

"I cannot wear the shield, because of the skin grafts across my face. The foam causes me to overheat."

Pause. Dead air on the phone line.

I said, "Listen, I'll accommodate you guys as well as I can. If you can find the mask that sits like a pair of glasses across the nose, I can wear that one for a small amount of time."

Pause. No sounds on the other line. Clicking of keyboard.

"Sir, we cannot find that mask you speak of."

"Then I cannot wear a mask."

"Then I'm afraid you cannot attend the event, sir."

"Do you know this decision you're making is against the law? This falls under the civil rights act. You are breaking the law."

"Sir, I'm sorry for the inconvenience, but we must comply with the CDC on this."

Okay, well, the CDC didn't ask you to do this. Not this way.

You know what's interesting? On my flight into New York City, I was sitting next to the wife of a US congressman. She recognized me, we chatted through the flight, and we exchanged numbers before we parted ways. I don't know if it was meant to be, but suddenly I had the phone number of a congressman and his wife. You can bet I texted them.

I also texted Jon Stewart, who was hosting the event.

Hey, Jon, it looks like I'm not coming to the event.

He picked up the phone and called me. "What do you mean, you're not coming?"

"Dude, they won't let me come unless I wear the mask. And I cannot wear the mask."

Jon said, "Then I'm not going either. If they won't let you in, then I'm out. I'll back out."

I have to be honest . . . I broke down when he said that. I was just so weary from fighting so hard all the time. It's hard enough to breathe, and it's harder still to continue to fight for all the people who are fighting that same fight.

"Oh, man, don't do that, Jon," I said. "Your solidarity and your sacrifice both mean a lot to me, but your presence means so much to the foundation and the veterans who are there. Don't do that for me."

If it was any other organization, I would have let him step down alongside me. But not the Woodruff Foundation. Not the veterans. They didn't deserve to be let down again.

I quietly decided I wasn't going to go to Stand Up for Heroes, my favorite event of the year, with my favorite nonprofit in the world.

Jon Stewart contacted Bob Woodruff, and Bob called me. "DT, you gotta go, man. You're one of my closest friends, and I want you there. But also, it means so much to the veterans that are here. When they see you, they are inspired. We want to give that to them tonight."

Always and forever, I'll do anything for the veterans. If we could find a mask I could wear, I'd wear the mask for as long as I could stand it. Also, I learned that the Bob Woodruff Foundation was behind me all the way. It was the venue making the rules.

My friend was with me, and we found a face shield. We shaved off the foam on the forehead piece, to see if I might not overheat as quickly. We got to the event, and we waited in line for about five minutes to get in, but I could already feel myself sweating. I knew I was only a few minutes away from overheating.

I looked at one of the security guards, and I asked him if I could pull the shield away from my face. Would that be acceptable if I just held it near my face? He gave me the thumbs-up. I thought, I'll wear it to get in, but when I sit down, I'm taking it off. I had word on good authority from the foundation. They were behind me the entire way.

We got inside, and someone from the Woodruff Foundation met me at the door. They said, "DT, we want to take you to the VIP area. The governor is there, some celebrities and other VIPs."

I followed them up to the exclusive room. I walked in to see women in ball gowns, men in tuxedos, waiters serving appetizers, an abundance of free liquor. And not a single person was wearing a mask.

I guess if you're a VIP, you don't have to wear a mask inside. That's bullshit. Not cool, you guys.

I took off my shield, and I tossed it in the nearest trash can, already mostly filled with cocktail napkins and appetizer plates.

By the end of that evening, the Bob Woodruff Foundation had raised $4.5 million. The night was a success, but the thoughtlessness of the venue made it difficult. You can't ignore the law, and you can't give special treatment to influential celebrities and rich people.

Listen, it's not just about me. What if there's a kid who wants to go see a play or a Broadway show, but they can't wear a mask because of their disability? Or what if it's their parent who cannot wear the mask, and so the kid can't go? You're telling disabled people not to come to plays, shows, entertainment, or even to New York City.

I am honored to be an advocate, to carry the shield

for those who cannot. But I have to be honest. Sometimes I just want to go to the theater, to the gym, to the grocery store, to the bank. Sometimes I wish somebody could make something easier for the people who have to work so hard to do even the easy things.

I decided to run for office, for reasons like this. People have encouraged me to run for a federal office, but right now, my heart is in the local influence. Always, always, always start with your local influence. Whether it's a charity or public service, I say you should always start local. The locals make more of an impact, and you want to have influence in the place where you live, where you can see your contribution at work.

I wanted to see our officials speaking for small businesses, for churches, for mom-and-pop shops. Small businesses had to shut down, but all the big-box stores were wide open. You couldn't go to church, but you sure could get your weed and alcohol.

I didn't see these local leaders leading by example. If they were going to mandate that these small businesses close their doors, then the leaders should have been willing to make the same sacrifice. They should have agreed not to take a paycheck until those small business owners could open their doors again. If they're going to make those decisions on behalf of the common people, then they should take the hit right along with the common folk.

That's why I decided to run for local office. I believe that our Founding Fathers didn't picture career politicians. I believe they thought we would have people in office to do their service for their country, and then they would go back to their lives and free the space and office to let the next person lead.

I have wondered sometimes if all of those difficult challenges were destined to prepare me for the greatest challenge of my life. I do believe that's the case, that "the big man upstairs" probably knew what was ahead of me, and he let all those other things happen to me to strengthen and prepare me.

So, I step in there. I get in there, and I lead. I made a promise to protect my brothers and sisters, and I can do that here and now—in my local community. This is the impact and influence I can have.

CHAPTER 26

I once heard a story about a college professor teaching a class on art appreciation. His students would look at a painting, a photo, or another visual art piece, and they would categorically decide, no. No, thanks. I don't like it.

So he taught them to reframe their thinking as they looked at the art. He taught them to ask this question: *If I liked something about this, what would I like?*

That question can change everything.

Back in 1997, I never thought I'd be this person. I thought I was going to be an active service member, doing my job, working with my teammates, and taking care of my family. Nobody thinks they're going to be injured, and then become a celebrity, where they know presidents and a prince. Nobody thinks that.

If you had told me decades ago that this would be the story of my life, I might have categorically decided, no. No, thanks. I don't like that plan.

Now that I'm here, looking at this life that is mine, I know this is my story. This is my opportunity to ask that same question the art professor asked his students: If there's anything I like about this, what do I like?

I like the people I've met. My teammates in basic training, my brothers downrange, the veterans I meet, each with the heart of a warrior.

I like that my son knows I'll fight for him, that the spark he gave me is what kept me alive.

I like that my wife is the best thing about me, my invincible other half. There is nothing she cannot do, and she's proved that a thousand times over. I like that I met her decades ago at a small-town festival in Mexico. In my heart, that woman is the queen of every pageant.

I like that I know how to motivate myself and others to keep going, even when it seems there's little left. My disability doesn't define me. How I choose to live my life is what defines me.

I like that the only ongoing physical discomfort I

have is nerve pain in my left hand. I wanted to get off the hard stuff right away, and I was willing to tolerate the pain so I didn't have to take heavy narcotics. If it weren't for my left hand, I probably wouldn't be on pain meds at all.

I like the teams of specialists who know what they're doing. I've had over 150 surgeries, as of today, and I know those surgeons, doctors, researchers, and biomedical engineers are preparing for the next person who needs their help. They're saving the world.

I like, because of my skin's sensitivity to the sun, that I'm legally allowed to have the windows of my truck tinted like a limousine.

I like that there's an ESPY in my house that belongs to me. The gold medals aren't bad, either.

Never in my life did I ever imagine that I would speak on a world stage, that the contacts in my phone would include Jon Stewart, George W. Bush, and Prince Harry.

I like going to sleep each night, knowing that I kept my promise to my dad. When my dad said, "Take care of your family," little did I know that this would include family members I would never meet in person. It wasn't just my brother and sisters, and it wasn't just my son. My dad may not have even known that that's what he meant. I like to imagine that I've made him proud.

There are parts of this story I would have chosen, and there are parts that I wouldn't have. But they all belong to me now. And I don't regret any of my choices.

Here's what I want to say to you, in these final moments together.

You and I both have a journey to take. We all do, and we can't see the path. When difficult things happen, we find ourselves always asking why. Sometimes we find that reason on our own, and sometimes we need someone else to help us find it.

Your support could come from family members or teammates, and sometimes even strangers. But here's what I'll tell you: don't be too proud to accept it. Take care of each other. Take care of your family. Because you and me—we are a family. We are in this together.

Carmen and I have discovered that there is always someone who has it worse than we do. When I was in the hospital, Carmen didn't know very much English, but she knew more than other people, so she translated for them. There was one woman who had raced out of the house when her family was injured, and she only had a T-shirt and pair of flip-flops. Carmen shared her clothes with her. When there's nothing you can do, you can give back. There's always someone you can help, and there's always a way that you can help. If you're struggling with your own problems, look up and see someone else. When you notice someone else, it takes your focus off yourself.

When I hear from someone who was having the hardest time, giving up on life, in their darkest hour, they say, "Sir, I was about to end it. And then I heard about you. I heard how you came out ahead. You helped me find my spark." That's when it hits me, what has happened here: all that pain, all that suffering, it was worth it. When I can help that one person, it gives me strength all over again. It's a cycle that continues, feeding itself with strength upon strength.

As I write this book, sixteen years have passed since the accident in Afghanistan, since the four months in a coma, since the long awakening to the life I have today. I probably could forget about the date on the calendar, except that it happened on Carmen's birthday. Every year that I celebrate the gift of her life, we remember what happened to us that day. We remember how far we have come, together.

To the young cadets I've mentored throughout the years, listen carefully. You don't become the best on your own, but you pick up pieces from those around you and those who have gone before you. That's how you get better, and it's how you take care of your teammates.

My team was there. My family was there for me. Every year, I get a call from Bailey, the one who lost his legs in that same accident. He calls me on the anniversary of the night before it all changed, when we watched the stars under the purest sky I had ever seen. He was eighteen when he lost his legs. He remembers with me. My teammates didn't let me be by myself. And in my recovery, I saw it as my mission to be there for my teammates. I continue to try to honor that promise I made to my dad, to take care of my brothers and sisters. And I raise that same challenge to you: take care of your brothers and sisters.

It still dumbfounds me that this is my life. I still can't believe people see me as an inspiration, as a motivation. I don't feel like anybody special. I did what any guy would have done in that same situation: I was there for my teammates, and I fought like hell. I only want to be known as a good teammate, a good husband, and a good father to my son. Everything else is extra. People like to call me a hero, but I'm just a dude who had a bad day at work.

As I write this, the Taliban has taken over Kabul,

Afghanistan, nearly twenty years after the September 11 attacks. I can tell you God's honest truth: no matter how I feel about the United States' exit, I would do every single bit of this over again. It all comes back to the promise I made to my dad, and I will forever do whatever it takes to keep my promise. I still live by the values my father gave to me: strength, faith, honor, and of course his words, "Always take care of your family."

Sometimes people think my family is the military—but my family includes civilians, too. Someone who was in a bad car accident. Someone who's struggling with bankruptcy or a failing business. Someone whose marriage is off the rails. When I promise to take care of my family, those words apply to anyone I know and even those I haven't met. We are in this together.

You're my family. You, the one reading this.

The healing comes in telling the story, and I'm telling my story to encourage you to keep going.

Stay Strong, Finish Strong, and Never Fucking Quit!

ACKNOWLEDGMENTS

It would not have been possible for me to write *A Patriot's Promise* without the help of many people.

Tricia Heyer is the collaborator who helped me bring my story to life on paper. Some people don't like to write, but some people cannot write. When I met Tricia, I held up my hands to show her my scars. I said, "I need someone to write my story. I don't have fingers anymore." Tricia said, "I'll take it from here, sir," and she has.

Dianne Derby is the Emmy Award–winning news anchor and journalist who has covered my story from every angle, and she continues to be a voice for veterans and Gold Star families in Colorado.

Greg Johnson is my literary agent.

St. Martin's Press is my publisher, and their team of editors and designers have done an outstanding job to bring this book to life.

Thank you to my doctors, nurses, and physical therapists, who are the unsung heroes in this journey.

To my family and friends, many of whom are noted in the book, thank you for having my six and, most importantly, being there for my wife and son.

My TACP brothers are the boots on the ground, doing the hard work of patriots and heroes.

Lastly, but most importantly, to my son, Israel, and my beautiful wife, Carmen:

Guero, you are my greatest accomplishment and you are my spark.

Gueris, you are the pillar of my strength and my heart.

INDEX